70

Advances in Biochemical Engineering / Biotechnology

Managing Editor: T. Scheper

Springer

Berlin
Heidelberg
New York
Barcelona
Hong Kong
London
Milan
Paris
Singapore
Tokyo

History of Modern Biotechnology II

Volume Editor: A. Fiechter

With contributions by
W. Beyeler, H. Bocker, H. R. Bungay,
E. DaPra, R. Katzen, W. A. Knorre,
N. W. F. Kossen, K. Schneider,
K. Schügerl, G. T. Tsao, J. M. Woodley

Springer

Advances in Biochemical Engineering/Biotechnology reviews actual trends in modern biotechnology. Its aim is to cover all aspects of this interdisciplinary technology where knowledge, methods and expertise are required for chemistry, biochemistry, microbiology, genetics, chemical engineering and computer science. Special volumes are dedicated to selected topics which focus on new biotechnological products and new processes for their synthesis and purification. They give the state-of-the-art of a topic in a comprehensive way thus being a valuable source for the next 3–5 years. It also discusses new discoveries and applications.

In general, special volumes are edited by well known guest editors. The managing editor and publisher will however always be pleased to receive suggestions and supplementary information. Manuscripts are accepted in English.

In references Advances in Biochemical Engineering/Biotechnology is abbreviated as Adv. Biochem. Engin./Biotechnol. as a journal.

ISSN 0724-6145
ISBN 3-540-67792-5
Springer-Verlag Berlin Heidelberg New York

Library of Congress Catalog Card Number 72-152360

Springer-Verlag Berlin Heidelberg New York
a member of BertelsmannSpringer Science+Business Media GmbH

Typesetting: Fotosatz-Service Köhler GmbH, Würzburg
Cover: E. Kirchner, Heidelberg
Printed on acid-free paper SPIN: 10753443 02/hu 3020 – 5 4 3 2 1 0

Managing Editor

Professor Dr. T. Scheper
Institute of Technical Chemistry
University of Hannover
Callinstraße 3
D-30167 Hannover/FRG
E-mail: scheper@mbox.iftc.uni-hannover.de

Volume Editor

Prof. Dr. A. Fiechter
Institute of Biotechnology
Eidgenössische Technische Hochschule Zürich
ETH-Hönggerberg
CH-8093 Zürich/Switzerland
E-mail: ae.fiechter@bluewin.ch

Editorial Board

Editorial

Over the last few years an alternation of generations in industry and the universities has taken place in Europe. Thus many of the leading biotechnologists in Europe who have been part of modern biotechnology during the last 30 years have already retired or will retire soon. The new and upcoming biotechnologists work on the basis of these efforts and often do not know much about the historic development after World War II, which brought us to the state of the art that we are now dealing with. When Prof. Dr. Armin Fiechter – one of these leading European biotechnologists – presented me with his idea of editing a special issue of the Advances in Biochemical Engineering/Biotechnology on the "History of Modern Biotechnology" I was extremely impressed to have the chance to review and summarize the historical development over the last 30 years. Prof. Fiechter is the first choice for this task, since he is the founder of the Advances in Biochemical Engineering/Biotechnology and not only did he help mold modern biotechnology but he has also been a continuous observer from the very beginning.

Prof. Fiechter succeeded in contacting biotechnologists from all over the world in order to summarize their points of view, especially in his own research areas, in different contributions. It was one of the main aims that personal views should also be included in the manuscripts in order to show how modern biotechnology was developed after World War II and how personal contacts, personal efforts, and personal opinion formed this research area. This generation of biotechnologists first succeeded in bringing together different areas of science to make this interdisciplinary research area into a powerful new technology. They had to overcome the hurdles existing between the different areas of science, especially biology, chemistry, engineering, mathematics and biochemistry and they had to build an international network to make biotechnology an international success.

These two special volumes (69 and 70) cannot be a complete detailed summing up of all biotechnological activities. However, these spotlights give a good overview. In particular this personal reviewing should give insights into the difficulties which had to be overcome and should give information about why certain decisions in the development of biotechnology were made. Our generation is sometimes confused as to why different developments were not speeded up or why it took so long to see that a certain direction in biotechnology was wrong. Several political and social obstacles are not well known any more. Thus, this special edition tries to give also an insight into these developments for a better understanding and act as a memorial to the scientists behind this development and their personal achievements in this success story called modern biotechnology.

I would like to thank all authors for helping Professor Fiechter to bring his idea to a successful fruition. It is their achievement that very different areas of biotechnology in different countries were brought together in a way to show the development of biotechnology in research, its industrial application and the personal and social involvement. I hope that these books will find a large number of young and older readers to present new insights into the roots of modern biotechnology.

Hannover, August 2000 Thomas Scheper

Preface

The aim of the Advances of Biochemical Engineering/Biotechnology is to keep the reader informed on the recent progress in the industrial application of biology. Genetical engineering, metabolism ond bioprocess development including analytics, automation and new software are the dominant fields of interest. Thereby progress made in microbiology, plant and animal cell culture has been reviewed for the last decade or so.

The Special Issue on the History of Biotechnology (splitted into Vol. 69 and 70) is an exception to the otherwise forward oriented editorial policy. It covers a time span of approximately fifty years and describes the changes from a time with rather characteristic features of empirical strategies to highly developed and specialized enterprises. Success of the present biotechnology still depends on substantial investment in R&D undertaken by private and public investors, researchers, and enterpreneurs. Also a number of new scientific and business oriented organisations aim at the promotion of science and technology and the transfer to active enterprises, capital raising, improvement of education and fostering international relationships. Most of these activities related to modern biotechnology did not exist immediately after the war. Scientists worked in small groups and an established science policy didn't exist.

This situation explains the long period of time from the detection of the antibiotic effect by Alexander Fleming in 1928 to the rat and mouse testing by Brian Chain and Howart Florey (1940). The following developments up to the production level were a real breakthrough not only biologically (penicillin was the first antibiotic) but also technically (first scaled-up microbial mass culture under sterile conditions). The antibiotic industry provided the processing strategies for strain improvement (selection of mutants) and the search for new strains (screening) as well as the technologies for the aseptic mass culture and downstream processing. The process can therefore be considered as one of the major developments of that time what gradually evolved into "Biotechnology" in the late 1960s. Reasons for the new name were the potential application of a "new" (molecular) biology with its "new" (molecular) genetics, the invention of electronic computing and information science. A fascinating time for all who were interested in modern Biotechnology.

True gene technology succeeded after the first gene transfer into *Escherichia coli* in 1973. About one decade of hard work and massive investments were necessary for reaching the market place with the first recombinant product. Since then gene transfer in microbes, animal and plant cells has become a well-established biological technology. The number of registered drugs for example may exceed some fifty by the year 2000.

During the last 25 years, several fundamental methods have been developed. Gene transfer in higher plants or vertebrates and sequencing of genes and entire genomes and even cloning of animals has become possible.

Some 15 microbes, including bakers yeast have been genetically identified. Even very large genomes with billions of sequences such as the human genome are being investigated. Thereby new methods of highest efficiency for sequencing, data processing, gene identification and interaction are available representing the basis of genomics – together with proteomics a new field of biotechnology.

However, the fast developments of genomics in particular did not have just positive effects in society. Anger and fear began. A dwindling acceptance of "Biotechnology" in medicine, agriculture, food and pharma production has become a political matter. New legislation has asked for restrictions in genome modifications of vertebrates, higher plants, production of genetically modified food, patenting of transgenic animals or sequenced parts of genomes. Also research has become hampered by strict rules on selection of programs, organisms, methods, technologies and on biosafety indoors and outdoors.

As a consequence process development and production processes are of a high standard which is maintained by extended computer applications for process control and production management. GMP procedures are now standard and prerequisites for the registation of pharmaceuticals. Biotechnology is a safe technology with a sound biological basis, a high-tech standard, and steadily improving efficiency. The ethical and social problems arising in agriculture and medicine are still controversial.

The authors of the Special Issue are scientists from the early days who are familiar with the fascinating history of modern biotechnology. They have successfully contributed to the development of their particular area of specialization and have laid down the sound basis of a fast expanding knowledge. They were confronted with the new constellation of combining biology with engineering. These fields emerged from different backgrounds and had to adapt to new methods and styles of collaboration.

The historical aspects of the fundamental problems of biology and engineering depict a fascinating story of stimulation, going astray, success, delay and satisfaction.

I would like to acknowledge the proposal of the managing editor and the publisher for planning this kind of publication. It is his hope that the material presented may stimulate the new generations of scientists into continuing the rewarding promises of biotechnology after the beginning of the new millenium.

Zürich, August 2000 Armin Fiechter

Contents

The Morphology of Filamentous Fungi
N. W. F. Kossen 1

Antibiotica Research in Jena from Penicillin and Nourseothricin to Interferon
H. Bocker, W. A. Knorre 35

Development of Bioreaction Engineering
K. Schügerl 41

A View of the History of Biochemical Engineering
R. Katzen, G. T. Tsao 77

Advances in Enzyme Technology – UK Contributions
J. M. Woodley 93

Computer Applications in Bioprocessing
H. R. Bungay 109

Automation of Industrial Bioprocesses
W. Beyeler, E. DaPra, K. Schneider 139

Author Index Volumes 51–70 165

Subject Index 173

Contents of Volume 69

History of Modern Biotechnology I

Volume Editor: A. Fiechter

The Natural Functions of Secondary Metabolites
A. L. Demain, A. Fang

Development of Applied Microbiology to Modern Biotechnology in Japan
T. Beppu

Microbial Production of Amino Acids in Japan
H. Kumagai

Development of Biotechnology in India
T. K. Ghose, V. S. Bisaria

History of Biotechnology in Austria
M. Roehr

Biotechnology in Hungary
J. Holló, U. P. Kralovánszky

Biotechnology in Switzerland and a Glance at Germany
A. Fiechter

The Morphology of Filamentous Fungi

N.W.F. Kossen

Park Berkenoord 15, 2641CW Pijnacker, The Netherlands
E-mail: kossen.nwf@inter.nl.net

The morphology of fungi has received attention from both pure and applied scientists. The subject is complicated, because many genes and physiological mechanisms are involved in the development of a particular morphological type: its morphogenesis. The contribution from pure physiologists is growing steadily as more and more details of the transport processes and the kinetics involved in the morphogenesis become known. A short survey of these results is presented.

Various mathematical models have been developed for the morphogenesis as such, but also for the direct relation between morphology and productivity – as production takes place only in a specific morphological type. The physiological basis for a number of these models varies from thorough to rather questionable. In some models, assumptions have been made that are in conflict with existing physiological know-how. Whether or not this is a problem depends on the purpose of the model and on its use for extrapolation. Parameter evaluation is another aspect that comes into play here.

The genetics behind morphogenesis is not yet very well developed, but needs to be given full attention because present models and practices are based almost entirely on the influence of environmental factors on morphology. This makes morphogenesis rather difficult to control, because environmental factors vary considerably during production as well as on scale. Genetically controlled morphogenesis might solve this problem.

Apart from a direct relation between morphology and productivity, there is an indirect relation between them, via the influence of morphology on transport phenomena in the bioreactor. The best way to study this relation is with viscosity as a separate contributing factor.

Keywords. Environmental factors, Filamentous fungi, Genetics, Modelling, Morphology, Physiology, Transport phenomena

1 General Introduction 3

2 The Framework of This Study 4

3 Introduction to Morphology 5

3.1 What Is Morphology 5
3.2 The Morphology of Filamentous Fungi 6

4 Overview of the Research 7

4.1 Methods 8
4.2 Models 10
4.2.1 Introduction 10

Advances in Biochemical Engineering/
Biotechnology, Vol. 70
Managing Editor: Th. Scheper

4.2.1.1 Building Blocks 11
4.2.1.2 Transport Mechanisms 11
4.2.1.3 Synthesis of the Cell Wall: Chitin 12
4.2.1.4 Synthesis of the Cell Wall: Glucan 13
4.2.1.5 Synthesis of the Cell Wall: the Structure 13
4.2.2 Morphology Modelling in General 14
4.2.3 Models for Morphogenesis 15
4.2.4 Models for the Relation Between Morphology and Production . . 20
4.2.5 Some General Remarks About Models 21
4.3 Special Aspects 24
4.3.1 Genetics 25
4.3.2 Whole Broth Properties 26

5 Implementation of the Results 28

6 Conclusions and Prospects 29

Appendix 30

References 32

List of Symbols and Abbreviations

C Concentration, kg m^{-3}
C_X Concentration of biomass, kg m^{-3}
DCR Diffusion with chemical reaction
ID Diffusion coefficient, m^2 s^{-1}
DOT Dissolved oxygen tension, N m^{-2}
D_r Stirrer diameter, m
d_h Diameter of hypha, m
ER Endoplasmatic reticulum (an internal structure element of a cell)
f(x, t) Population density function: number per m^3 with property x at time t
k_1, k_2 Lumped parameters
$k_l a$ Mass transfer parameter, s^{-1}
L Length of hypha, m
L_e Length of main hypha in hyphal element, m
L_{emax} Maximum length of main hypha capable of withstanding fragmentation, m
L_{equil} Equilibrium length, m
L_t Length of all hyphae in hyphal element, m
L_{hgu} Length of hyphal growth unit (L_t/n), m
m mass, kg
m_{hgu} Mass of a hyphal growth unit, kg per tip
N Rotational speed of stirrer, s^{-1}
n Number of tips in hyphal element, -

NADP Nicotinamide adenine dinucleotide phosphate: oxydation/reduction coenzyme in which NADPH is the reducing substance
P/V Power per unit volume of fermenter, W m^{-3}
r Distance to stirrer, m
r (C) Reaction rate as function of C, kg m^{-3} s^{-1}
r_l Rate of vesicle production per unit length of hypha, number m^{-1} $s^{-1)}$
Rho 1p A GTP-binding enzyme involved in the cell awl synthesis
t_c Circulation time, s
V Volume, m^3
v Velocity, m s^{-1}
V_{disp} Volume with maximum dispersion potential, m^3
z Vector representing the environmental conditions, varying dimensions
ε Power per unit mass, W kg^{-1}
ϕ_p Pumping capacity of stirrer, m^3 s^{-1}
γ Shear rate, s^{-1}
μ Specific growth rate, s^{-1}
τ Shear stress, N m^{-2}

1 General Introduction

Filamentous fungi are fascinating organisms, not only because of the inherent beauty of their fruiting bodies but also because of their complicated and scientifically very interesting behaviour. They are also able to produce a large variety of useful , commercially interesting products.

The use of filamentous fungi as production organisms in industry, originally as surface cultures, is widespread,. Many scientists once believed that these fungi could only grow as surface cultures but it became clear in the 1940s that submerged cultures are also possible and have an enormous production potential. However, there appeared to be one problem: their form. In their natural environment filamentous fungi grow in long, branched threads called hyphae. This form, which is ideal for survival in nature, presents no problem in surface cultures, but it is often a nuisance in submerged cultures because of the strong interaction between submerged hyphae. This results in high apparent viscosities ("applesauce" behaviour) and – as a consequence – in major problems in the transport of O_2, CO_2, and nutrients, as well as in low productivities compared with theoretical values and with productivities obtained with other microorganisms. It was obvious that the control of the form of these fungi was a real issue that needed further attention in order to make optimal use of their potential production capacities.

Many scientists have been studying this problem from an engineering point of view for a number of decades. Simultaneously, many other scientists, working on morphology mainly because of pure scientific interest or sheer curiosity, have been very active.

The outcome of the efforts mentioned above is an impressive landscape of results about what is now called "the morphology of fungi". This paper is about this landscape: what it looks like, how it emerged and developed, which tools were developed, and what are its strengths and weaknesses.

2
The Framework of This Study

As will be clear from the introduction this is not another review on the morphology of fungi. There are excellent, up-to-date and extensive reviews available [1]. This is a survey of the main lines of development of a very interesting area of biotechnology research. based on a limited number of characteristic publications. These have been selected on the basis of their contributions – either good or debatable ones – to new developments in two areas:

- Improved scientific insight.
- Bioprocess practice – is it useful and usable?

The improvement of scientific insight usually goes hand in hand with a number of developments in the models used (see Fig. 1). These developments provide the main yardsticks for the present evaluation.

The trend in the development from unstructured to structured models needs an introduction. In unstructured models one assumes that the object of study has no structure: for example, a hyphal element is considered to be a more-or-less black box without internal detail. If one distinguishes septa, nuclei etc., the model then becomes structured. This structuring can go on a long way and become very detailed, but a limited number of internal "compartments" is usually sufficient to describe an observed phenomenon properly.

In the literature, models of another useful kind are sometimes mentioned: segregated – or corpuscular – models. In that case, a population is not considered to be a unit with average properties, but a collection of different individuals, each with its own properties: form, size, respiration rate, etc.

The methods used for the parameter optimization and the validation of the models will also be part of the evaluation.

Three classes of subjects will be discussed:

1. Methods: image analysis, microelectrodes, single hyphal elements, staining.
2. Models: models for morphogenesis and for the relation between morphology and production.
3. Special aspects: genetics, transport phenomena.

Now that the subjects and the yardsticks have been presented, just one word about the the author's viewpoint. This point of view is that of a former uni-

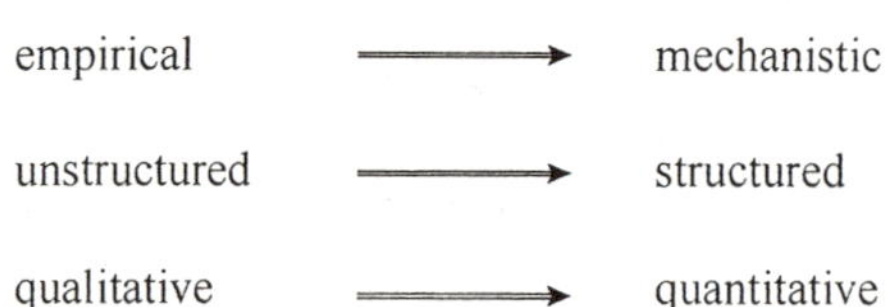

Fig. 1. Development of models

versity professor, who started research after the morphology of moulds in 1971 and – inspired by problems he met as a consultant of Gist-brocades – worked in this particular area of biotechnology for about 10 years. After 17 years at the university, he went to Gist-brocades and worked there for 10 years. Most of the time as a director of R&D, in which position he became heavily involved with technology transfer among all of the disciplines necessary for the development of new products/processes and the improvement of existing ones.

3 Introduction to Morphology

3.1 What Is Morphology?

Morphology is the science of the form of things. It is a wide spread field of attention in a large number of sciences: biology as a whole, geology, crystallography, meteorology, chemistry – biochemistry in particular, etc. It usually starts as a way of classifying objects on the basis of their form. When the scientist becomes curious about the "why" of the development of a form he/she gets involved in the relationship between form and function. In the end, this can result in the prediction of properties given a particular form, or in the control of form/function.

First, we need several definitions. A hypha (plural: hyphae) is a single thread of a hyphal element. A hyphal element consists of a main hypha, usually with a number of branches, branches of branches etc., that originates from one spore. A flock is a loosely packed, temporary agglomerate of hyphal elements. A pellet or layer is a dense and -- under normal process conditions – almost permanent configuration of hyphae or hyphal elements (see Fig. 2).

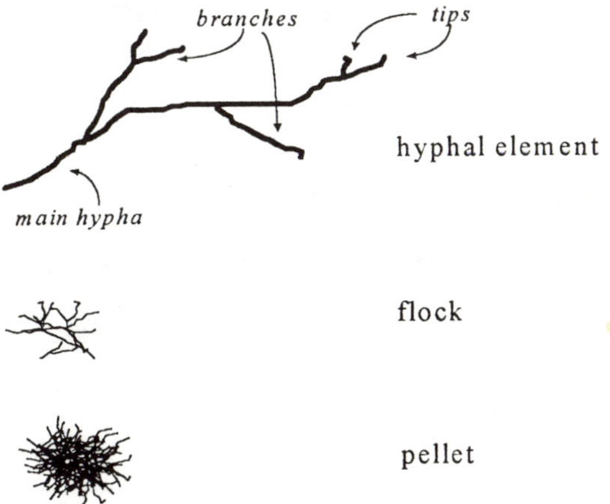

Fig. 2. Several definitions and forms

Furthermore, the "form of things" is a rather vague concept that needs further specification. The morphology of fungi is usually characterized by a limited number of variables, all related to one hyphal element: the length of the main hypha (L_e), the total length of all the hyphae (L_t), the number of tips (n) and the length of a hyphal growth unit (L_{hgu}). The L_{hgu} is defined as L_t/n.

3.2 The Morphology of Filamentous Fungi

The various forms of filamentous fungi have advantages and disadvantages in production processes as regards mass transport properties and the related overall (macro) kinetics, in particular at concentrations above 10–20 kg m^{-3} dry mass (see Table 1). As has already been mentioned, the poor transport properties are the result of the strong interaction between the single hyphal elements at high biomass concentrations, often resulting in fluids with a pronounced structure and a corresponding yield stress. This results in poor mixing in areas with low shear and in bad transport properties in general.

Morphology is strongly influenced by a number of environmental conditions, i.e. local conditions in the reactor:

1. Chemical conditions like: C_{O_2}, $C_{substrate}$, pH.
2. Physical conditions like: shear, temperature, pressure.

We will use the same notation as Nielsen and Villadsen [2] to represent all these conditions by one vector (**z**).Thus morphology(**z**) means that the morphology is a function of a collection of environmental conditions represented by the vector **z**. If necessary **z** will be specified.

Also, genetics must have a strong influence on the morphology, because the "genetic blueprint" determines how environmental conditions will influence morphology. We will return to this important issue later on. For the time being, it suffices to say that at present, despite impressive amounts of research in this area, very little is known that gives a clue to the solution of production problems due to viscosity in mould processes. This situation shows strong similarity with the following issue.

Table 1. Transport properties of various forms of moulds

Form of element	Transport to element within broth	Transport within element	Mechanical strength of element
Single hyphal elements	–/+[a]	+	±
Flocs	–/++[b]	±	–
Pellet/layer	+	–	+

[a] Depending on the shape, size and flexibility of the hyphal element.
[b] Depending on kinetics of floc formation and rupture.

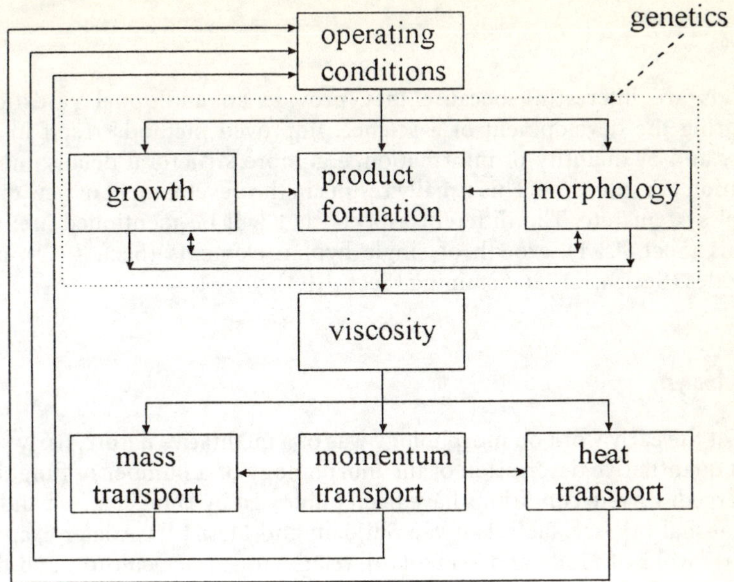

Fig. 3. Mutual influences between morphology and other properties

A very important practical aspect of the morphology of filamentous fungi is the intimate mutual relationship between morphology and a number of other aspects of the bioprocess. This has already been mentioned by Metz et al. [3], in the publication on which Fig. 3 is based. The essential difference is the inclusion of the influence of genetics. In this figure, viscosity is positioned as the central intermediate between morphology and transport phenomena. Arguments in support of a different approach are presented in Sect. 4.3.2.

This close relationship, which – apart from genetics to some extent – is without any "hierarchy", makes it very difficult to master the process as a whole on the basis of quantitative mechanistic models. The experience of the scientists and the operators involved is still invaluable; in other words: empiricism is still flourishing.

Morphology influences product formation, not only via transport properties – as suggested by Fig. 3 – but can also exert its influence directly. Formation of products by fungi can be localized – or may be optimal.– in hyphae with a specific morphology, as has been observed by Megee et al. [4], Paul and Thomas [5], Bellgardt [6] and many others.

4
Overview of the Research

This chapter comprises three topics: methods, models, aspects.

4.1 Methods

Methods are interesting because they provide an additional yardstick for measuring the development of a science. Improved methods result in better quality and/or quantity of information, e.g. more structural details, more information per unit time. This usually results in the development of new models, control systems etc. The different aspects that will be mentioned are: image analysis (Sect. 4.1.1), growth of single hyphal elements (Sect. 4.1.2), microelectrodes (Sect. 4.1.3) and staining (Sect. 4.1.4).

4.1.1 *Image Analysis*

Much of the early work on morphology was of a qualitative nature. Early papers with a quantitative description of the morphology of a number of fungi under submerged, stirred, conditions have been published by Dion et al. [7] and Dion and Kaushal [8] (see Table 1 of van Suijdam and Metz [9]). A later example is the early work of Fiddy and Trinci [10], related to surface cultures and that of Prosser and Trinci [11]. Measurements were performed under a microscope, by either direct observation or photography. The work can be characterized as extremely laborious.

In their work, Metz [12] and Metz et al. [13] made use of photographs of fungi, a digitizing table and a computer for the quantitative analysis of the above-mentioned morphological properties of filamentous fungi (L_e, L_t, n and L_{hgu}) plus a few more. Although the image analysis was digitized, it was far from fully automated. Therefore, the work was still laborious, but to a lesser extend than the work of the other authors mentioned above.

The real breakthrough came when automated digital *i*mage analysis (ADIA) was developed and introduced by Adams and Thomas [14]. They showed that the speed of measurement – including all necessary actions – was greater than the digitizing table method by about a factor 5. A technician can now routinely measure 200 particles per hour. Most of the time is needed for the selection of free particles.

Since then, ADIA has been improved considerably by Paul and Thomas [15]. These improvements allow the measurement of internal structure elements, e.g. vacuoles [16], and the staining of parts of the hyphae, in order to differentiate various physiological states of the hyphae by Pons and Vivier [17].

Although the speed and accuracy of the measurements, as well as the amount of detail obtained, show an impressive increase, there are areas , e.g. models, where improvement of ADIA is essential for further exploration and implementation. An important area is the experimental verification of population balance, in which case the distribution in a population of more than 10,000 elements has to be measured routinely [18]. This is not yet possible, hampering the verification of these models. For average-property models, where only average properties have to be measured, 100 elements per sample are sufficient, and this can be done well with state-of-the-art ADIA.

Closely related to ADIA is automated sampling, which allows on-line sampling and measurement of many interesting properties, including morphology. This method is feasible but is not yet fast and accurate enough [17].

Needless to say, in all methods great care must be taken in the preparation of proper samples for the ADIA. Let this section end with a quotation from the thesis of Metz [12] (p. 37) without further comment. It reads: "The method for quantitative representation of the morphology proved to be very useful. About 60 particles per hour could be quantified. A great advantage of the method was that the dimensions of the particles were punched on paper tape, so automatic data analysis was possible".

4.1.2 *Growth of Single Hyphal Elements*

Measurement of the growth of single hyphal elements is important for understanding what is going on during the morphological development of mycelia. It allows careful observation , not only of the hyphae such as hyphal growth rate, rate of branching etc., but also – to some extent – of the development of microstructures inside the hyphae, such as nuclei and septa. This has contributed considerably to the development of structured models. There are early examples of this method [10], in which a number of hyphal elements fixed in a surface culture were observed. An example of present work in this area has been presented by Spohr [19]. A hyphal element was fixed with poly-L-lysine in a flow-through chamber. This allows for the measurement of the influence of substrate conditions on the kinetics of morphological change in a steady-state continuous culture with one hyphal element. This work will be mentioned again in Sect. 4.2.3.

4.1.3 *Staining*

Another technique that has contributed to the structuring of models is the use of staining. This has a very long history in microbiology, e.g. the Gram stain, in which cationic dyes such as safranin, methylene blue, and crystal violet were mainly used. Nowadays, new fluorescent dyes and/or immuno-labelled compounds are also being used [17, 20], allowing observation of the internal structure of the hyphae. A few examples are listed in Table 2:

Table 2. Staining

Dye	What does it show?
Neutral red	Apical segments
Methylene blue/Ziehl fuchsin	Physiological states in *P. chrysogenum*
Acridine orange (AO) *fluoresc.*	RNA/DNA (single or double stranded)
Bromodeoxyuridine (brdu) *fluoresc.*	Replicating DNA
Neutral red	Empty zones of the hyphae
Methylene blue/Ziehl fuchsin	

Applications in morphology have been mentioned [17, 20]. Several examples are:

- Distinction between dormant and germinating spores; location of regions within hyphae – as well as in pellets – with or without protein synthesis (AO).
- Propagation in hyphal elements (BrdU) in combination with fluorescent antibodies).
- These techniques contribute to the setup and validation of structured models.
- Measurement of NAD(P)H-dependent culture fluorescence, e.g. for state estimation or process pattern recognition, is also possible [21].

4.1.4
Micro-Electrodes

As has already been mentioned in Sect. 3a (Table 1), filamentous fungi, among others, can occur as pellets or as a layer on a support. This has both advantages and disadvantages. An example of the latter is limitation of mass transfer and, therefore, a decrease in conversion rate within the pellet or layer compared with the free mycelium. The traditional chemical engineering literature had developed mathematical models for this situation long before biotechnology came into existence [22] and these models have been successfully applied by a whole generation of biotechnologists. The development of microelectrodes for oxygen [23], allowing detailed measurements of oxygen concentrations at every position within pellets or layers, opened the way to check these models. Hooijmans [24] used this technique to measure the O_2 profiles in agarose pellets containing an immobilized enzyme or bacteria. Microelectrodes have also been used to measure concentration profiles of O_2 and glucose (Cronenberg et al. [25]) as well as pH and O_2 profiles [26] in pellets of *Penicillium chrysogenum*. These measurements were combined with staining techniques (AO staining and BrdU immunoassay). This resulted in interesting conclusions regarding a number of physiological processes in the pellet.

Much of what has been mentioned above about methods , such as staining and microelectrodes, has been combined in Schügerl's review [20]. This publication also discusses a number of phenomenological aspects of the influence of environmental conditions (z), including process variables, on morphology and enzyme production in filamentous fungi, mainly *Aspergillus awamori*.

4.2
Models

4.2.1
Introduction

A majority of the models describing the morphogenesis of filamentous fungi deal with growth and fragmentation of the hyphal elements. Structured models have been used from early on. A number of them will be shown in this paragraph, but some physiological mechanisms of cell wall formation are presented first

The basis for mechanistic, structured, mathematical models describing the influence of growth on the morphogenesis of fungi is physiology. At least, the basic assumptions of the model should not contradict the physiological facts. Therefore, a brief overview of the physiology of growth, based mainly on a publication of Gooday et al. [27], is presented here. Emphasis is on growth of Ascomycetes and Basidiomycetes, comprising *Penicillium* and *Aspergillus*, inter alia. In other fungi, the situation may be different.

Growth of fungi manifests itself as elongation – including branching – of the hyphae, comprising extension of both wall and cytoplasm with all of its structural elements: nuclei, ER, mitochondria and other organelles. The morphology of fungi is determined largely by the rigid cell wall [28]; therefore, this introduction is limited to cell-wall synthesis.

Cell-wall synthesis in hyphae is highly polarized, because it occurs almost exclusively at the very tip, the apex.

4.2.1.1
Building Blocks

The major components of the cell wall are chitin and glucan. Chitin forms microfibrils and glucan the matrix material in between them. The resulting structure is very similar to glass-fiber reinforced plastic.

Vesicles, containing precursors for cell wall components and enzymes for synthesis and transformation of wall materials, are formed at the endoplasmatic reticulum (ER), along the length of the active part of the hyphae. The concentration of vesicles in the hyphal compartment increases gradually from base to tip by about 5% by volume at the base, to 10% at the tip, with the exception of the very tip, where a rapid increase in the vesicle concentration is observed. At that point, up to 80% by volume of the cytoplasm may consist of vesicles.

4.2.1.2
Transport Mechanisms

This subject deserves some attention, because it is a common mechanism in all polarized growth models. Vesicles are transported to the tip by mechanisms that are still obscure. A number of suggestions for this transport mechanism have been summarized [27]

1. Electrophoresis due to electropotential gradients.
2. A decline in concentration of K^+ pumps towards the tip, resulting in a stationary gradient of osmotic bulk flow of liquids and vesicles to the tip.
3. A flow of water towards the tip, due to a hydrostatic pressure difference within the mycelium.
4. Cytoplasmic microtubules guiding the vesicles to the tip.
5. Microfilaments involved in intracellular movement.

Diffusion is excluded from this summary because the concentration gradient towards the tip increases (i.e. $dC_{vesicles}/dx > 0$), and therefore passive diffusion cannot play a role.

With regard to point 3, microscopically visible streaming of the cytoplasm is said to occur in fungi [29]. It is likely, however, that what has been observed is not the flow as such, but the movement of organelles. The two cannot be distinguished, because we are unable to perceive movement without visual inhomogeneities, such as particles, bubbles, clouds, etc. Moreover, the mechanism behind this movement does not have to be flow. The presence of flow is not likely, because flow needs a source and a sink. The source is present, i.e. uptake of materials through the cell membrane, but where is the sink? A sink could be withdrawal of materials needed for extension of the hyphae, but then the flow would never reach the very tip. Recirculation of the flow could be a solution for the source/sink problem but then a "pump" is needed, and it is not clear how this could be realized. Therefore transport to the apex is difficult to envisage. Passive diffusion is not possible, because the concentration gradient is positive, and flow within the cytoplasm is unlikely, because there is no sink.

An interesting hypothesis, that is an elaboration of the points 4 and 5, has recently been suggested, which can solve the problems of diffusion and flow. Howard and Aist [30] and others have shown that cytoplasmic microtubules play an important role in vesicle transport, because a reduction in the number of microtubules in *Fusarium acuminatum* inhibits vesicle transport. Regalado et al. [31] have given a possible explanation of the transport of vesicles, based on the role of microtubules and the cytoskeleton in general. They consider two transport mechanisms, diffusion and flow. In particular, their proposal for the diffusion process has a very plausible basis. In the literature, the usual driving force for transport by diffusion is a concentration gradient, but they propose a different mechanism. If stresses of a visco-elastic nature are applied to the cytoskeleton, the resulting forces are transmitted to the vesicles. The vesicles experience a force gradient that converts their random movement in the cytoskeleton to a biased one. Consequently, they move from regions of high stress to regions of low stress. The driving force is thus no longer a concentration gradient but a stress gradient, thus solving the problem that arose with the classical diffusion. Their computer simulations look very convincing, but more experimental evidence is needed. They cite many other examples from the literature showing the relation between cytoskeletal components and vesicle transport.

4.2.1.3
Synthesis of the Cell Wall: Chitin

Chitin synthesis occurs exclusively at the growing hyphal tip and wherever cross-walls (septa) are formed. This indicates that chitin synthesis has to be closely regulated both in space and time. All of the genes for chitin synthase that have been isolated so far code for a protein with an N-terminal signal sequence. This indicates that the protein is synthesized at the ER, transported through the Golgi and brought to the site of action, i.e. the hyphal tip or the site of cross-wall formation, in secretory vesicles. There, it functions as a transmembrane protein, accepting the precursor UDP-N-acetyl-glucosamine at the cytoplasmic site and producing chitin polymers on the outside. There, in the wall area, different chitin polymers interact by mutual H-bonding, and crystallize spontaneously into microfibrils. This crystallization process might be hampered by the cross-linking of newly synthesized chitin to other wall components.

Early findings for many fungal chitin synthases were that these enzymes are often isolated in an inactive state and can be activated by proteolytic digestion [32]. This led to the idea that synthetases may be regulated by transformation of a zymogen form into an active enzyme, but not necessarily by proteolysis. Treatment of membrane preparations with detergent resulted in loss of activity that could be restored by addition of certain phospholipids, indicating that the lipid environment in the membrane might be another possible activating factor.

4.2.1.4
Synthesis of the Cell Wall: Glucan

The enzyme involved in glucan synthesis is also a membrane bound protein that catalyses the transfer of glucosyl residues from UDP-glucose to a growing chain of β-1,3- linked glucosyl residues. It was found that the synthase is highly stimulated by micromolecular concentrations of GTP. Subsequently, testing mutants for GTP-binding proteins with a phenotype compatible with a defect in cell wall synthesis, *Rho 1*-mutants were found. Tests of these mutants established unequivocally that Rho 1 protein (Rho 1p) is the GTP-binding protein that regulates β-1,3-glucan synthase and is essential for its activity [33].

Intriguingly, Rho 1p has two further functions: it regulates both cell-wall synthesis and morphogenesis. Rho 1p activates protein kinase C, which in turn regulates a pathway that leads to cell-wall synthesis in response to osmotic shock [34]. However, Rho 1p is not directly involved in the activation of β-1,3-glucan synthesis. Furthermore, Rho 1p may also be involved in the organization of the actin cytoskeleton at the hyphal tip (Yamochi et al. [35]).

4.2.1.5
Synthesis of the Cell Wall: the Structure

Finally, covalent bonds are formed between chitin and glucan polymers, and hydrogen bonds are formed between the homologous polymer chains, resulting in a strong combination of chitin fibers in a glucan matrix (Wessels [36]). Wessels also makes it clear that the spatial distribution of wall extrusion and progressive crosslinking might result in different morphologies: mycelium, pseudo-mycelium, and yeast.

There are additional compounds present in the membrane and other enzymes are also involved, but the essentials needed for the evaluation of the growth models in this paragraph have been presented above.

It should be clear that the growth of the cell wall is a rather complicated process, with many enzymes involved and much that is still unknown, but a number of facts are known that can exclude certain mechanisms.

Before we start the discussion about mathematical growth models, two more general remarks regarding structured models have to be made. Structured models are not necessarily of a mechanistic nature. One can observe one hyphal element under the microscope and describe all the internal structures one sees exactly, without any mechanistic explanation. This description can also be quantitative but it remains empirical (*the "flora" is a typical example of a non-mechanistic structured approach: it shows all the different organisms in a habitat – and often their parts as well – in a systematic way, without explaining "why"*).

The models that we will discuss have been validated by experiments with various fungi. However, the morphological characteristics of moulds vary enormously between different strains, even between strains belonging to the same species. In other words, the quantitative results are very specific. Therefore, the main values of a well-validated model are the methodology and the structure, not the actual figures.

4.2.2 *Morphology Modelling in General*

A systematic survey of the modelling of the mycelium morphology of *Penicillium* species in submerged cultures has recently been published [18]. The authors distinguish between various of kinds of models (see Fig. 4). A short explanation follows.

Models of Single Hyphal Element. Experiments with single hyphal elements were mentioned in Sect. 4.1.2. The advantages, greater detail and more insight, can be used in single hyphal element models. Several examples have been mentioned [11, 18].

Population Models. In population models, or population balances, microorganisms are treated as individuals with different properties. Each individual hyphal element has its own properties, in this case usually L_t and n. Central in these models is the population density function, $f(L_t, n, t)$, representing the number of individuals per volume in the population with a specific value of L_t and n at time t. The value of $f(L_t, n, t)$ can change under the influence of tip

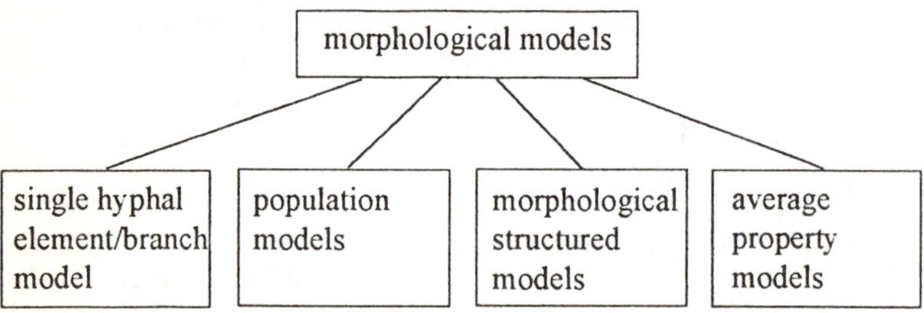

Fig. 4. Kinds of morphological models

extension, branching, birth of new hyphal elements due to germination or fragmentation, and – in continuous cultures – dilution.

Population balances were used in the biotechnology of bacteria and yeasts, before the term had been coined (e.g. [37]), but not for filamentous fungi. The main reason is that the verification of these models required rapid methods for measuring the properties of the individuals. This is no problem for individuals of the size of bacteria and yeasts Thousands of cells can be measured quickly and routinely (Coulter counter). For filamentous fungi the situation is different. For the characterization of the morphology of fungi, not one but at least two variables are important, e.g. L_t and n. As mentioned in Sect. 4.1.1, they have to be measured for so many elements that it is too much even for ADIA at present. Therefore, we will not deal with these models.

Nevertheless, these models are very elegant and allow for very detailed description of microbial systems [18]. One thing is certain, the processing and storage capacity of computers will increase drastically so that the prospects for use of these models in the future are favourable.

For those interested in the background of population balances, two publications are recommended: Randolph [38] and Randolf and Larssen [39].

Morphologically Structured Models. These models deal with conversions between different morphological forms, resulting in shifts in fractional concentrations of biomass with a specific morphological form (Nielsen [40]).

Average Property Models. These models deal with the averages of the population as a whole, i.e. average length, average number of tips per hyphal element (e.g. [12, Aynsley et al. 41, Bergter 42]).This group forms the vast majority of all models – not only in morphogenesis!

We will deal with models for the development of a particular morphology (morphogenesis) and models that include the influence of the morphology on the production of metabolites – usually antibiotics.

4.2.3
Models for Morphogenesis

These models contain only mechanisms for growth, branching and fragmentation. In fact, these "single" mechanisms are usually the result of underlying submechanisms, and so on. One well-known example has already been mentioned [11]. The only mechanism is growth but , as we will see below, there are seven or eight submechanisms, depending on the degree of subdivision.

Morphogenesis is the development of a particular morphological form. For fungi this form is usually characterized by the number of tips and either the total hyphal length (L_t) or the length of the main hypha (L_e) (both per hyphal element).

The first models dealt only with growth. Examples are the constant linear extension rate of pellets and hyphal colonies, apart from the very beginning of hyphal growth (Trinci [43, 44]), the branching after a certain length of the hypha (the first structuring (Plomley [45])), and the constant value of the

L_{hgu} (Caldwell and Trinci [46]). The L_{hgu} is not constant if the diameter of the hypha varies [40]. Much more constant is the mass of a hyphal growth unit (m_{hgu}) where $m_{hgu} = m/n_{tot}$.

Because branches also grow at a constant rate, form new branches etc., this results in exponential growth of the mycelium under non-limiting substrate conditions.

One of the other older growth models worth mentioning is the cube-root law for the growth of pellets (Emerson [47]). It was presented as an empirical model, but the mechanism behind it is obvious once one realizes that, under the condition of substrate limitation, growth of a pellet can only occur in a thin layer of almost constant thickness. The amount of structuring in these models is rather low or absent.

The study of single hyphal elements [19] revealed that the tip extension rate of the hypha can be modelled with saturation kinetics with respect to the branch length. Nielsen and Krabben [48] found the same relation for the average tip-extension rate when samples of about 50 hyphal elements were measured.

The next group of models are representatives of morphogenesis as a result of growth and fragmentation.

The earliest work on fragmentation was performed at the Istituto Superiore di Sanità in Rome [7, 8] where extensive phenomenological research has been done on the influence of stirring on the morphology of *Penicillium chrysogenum* and a number of other filamentous fungi. They noticed a strong correlation between morphogenesis and stirrer speed, but did not correlate their results in mathematical form. This was done by others [12] and has also been published in a review [9]. The dependence of the effective length (L_e) on P/V could be represented by $L_e = \text{const.}\ \varepsilon^{\alpha}$, where ε represents the power per unit mass of the broth. The value of α varies from -0.66 to -1.11.

Growth and fragmentation of *Penicillium chrysogenum* were combined in one model rather early [12, 9]. The growth model used was based on early work [45, 46]. Hyphae grow at a constant linear rate. After $\Delta L = L_{hgu}$, a new branch is formed, etc. This results in exponential growth of the total length per hyphal element, and of the culture as a whole. The only structure included in the growth model is branching. The model used for fragmentation is based on the turbulence theory of Kolmogorov (see [49]). Breakup of the main hypha is due to eddies formed as a result of local energy input into the medium. First, the maximum effective length of the hyphal element $L_{e_{max}}$ before fragmentation occurs is calculated. $L_{e_{max}}$ is the maximum length of the main hypha):

$$L_{e_{max}} = const. \frac{d_h^{0.38}}{\varepsilon_{max}^{0.25}} \tag{1}$$

The constant includes, inter alia, wall thickness and tensile strength of hyphae; d = diameter of hyphae.

The main hypha will break up only if $L_e > L_{e_{max}}$. This will only occur in a limited volume of the reactor near the stirrer (V_{disp}), where ε is at a maximum.

The dynamics of the fragmentation process is introduced by assuming that its rate is determined by the probability of fragmentation, which is proportional to the number of times the element passes V_{disp}.

V_{disp} is a function of ε, N, D_r and the distance to the stirrer (r). The final result is:

$$-\frac{dL_e}{dt} = const.\, N^{1.75} \cdot D_R^{0.5} \cdot d^{-0.38} \cdot (L_e^2 - L_{e_{\max}}^2) \tag{2}$$

for $L_e \geq L_{e_{\max}}$, otherwise $dL_e/dt = 0$.

This is one of the first detailed mechanistic models for the fragmentation of hyphal elements. Among other factors, it takes into account the fact that the turbulent energy is absorbed in a small region near the stirrer, so cannot be considered to be evenly distributed. As we can see, dL_e/dt is proportional to L_e^2 Other authors assume a different dependence on L_e.

The steady-state value of L_e can be calculated assuming that the growth rate is equal to the fragmentation rate. The approximate result is:

$$L_e = k_1 d^{0.38} / N^{0.75} D_R^{0.5} + k_2 \mu / N \tag{3}$$

The growth and fragmentation of pellets has been studied by van Suijdam [50]. The model of van Suijdam includes growth and autolysis, mass transfer of oxygen – described as density-dependent diffusion accompanied by chemical reaction, as well as external oxygen transfer from bubble to liquid and from liquid to pellet. The basis for the fragmentation is again the Kolmogorov theory [51]. The mechanistic model for fragmentation of the pellets can be summarized as follows: $d_{aver} = c\varepsilon^{-0.4}$, where c is a constant; the experimentally determined power of ε was 0.38.

To obtain a model for spore germination, growth and fragmentation of *Penicillium chrysogenum* a population balance was used as a start to derive balances for average properties [48]. By using the results of their own experiments and experimental data from the literature, it was possible to extract interesting information about germination, growth and fragmentation of hyphal elements. A very good fit was obtained between model and experiments for those periods during the bioprocess in which the conditions for the extraction were fulfilled, e.g. no fragmentation during germination.

Ayazi Shamlou et al. [52] present a fragmentation model with first-order dependence in L_e:

$$\frac{dL_e}{dt} = k \cdot (L_{e_{equil}} - L_e) \tag{4}$$

They too use the Kolmogorov theory to calculate L_{equil} but they assume that the turbulence in the vessel is homogeneous and isotropic, which is not correct.

The models mentioned above have a very limited physiological background. This is particularly evident in the case of the fragmentation models, although there certainly are opportunities to include physiological mechanisms in fragmentation, for example, vacuolization and subsequent decrease in the local tensile strength of the hyphae. The growth models are also almost non-physiology based; only L_{hgu} has a physiological background.

We will now deal with a number of physiology-based models.

An often quoted example of an early mechanistic and structured growth model of a single hyphal element, based on solid physiological observations [11], is an extension of earlier work of the same group [10]. The structural elements are: tips, septa and the regions between them, vesicles and nuclei. Growth – in fact, cell-wall synthesis – occurs at the tips of the hyphae as a result of the *inclusion of vesicles*. These *vesicles are produced* at a constant rate at the wall of the hyphae, *transported* to the tips and used for growth. Under the influence of growth, the ratio of cytoplasmic volume and the number of nuclei at the tips in the apical compartment attains a critical value. This triggers *an increase in the number of nuclei* until they have doubled (from four to eight). These eight nuclei are then divided into two sections, each of four nuclei, by *the formation of a new septum* between them, and so on. These septa have an *opening, the diameter decreases with age*, i.e. with distance from the tip. This results in a *congestion of vesicles* before such a septum, which – in turn – triggers *branch formation*. The submechanisms have been *underscored*.

Several other attempts have been made to increase the physiological basis, or the structuring, of the growth mechanism in models. An example [41] is the use as a growth model of a self-extending tubular reactor that absorbs nutrients along its length. One of the assumptions is transport of the precursor through the hyphae at a constant average rate of flow. As has already been mentioned in Sect. 4.2.1, this is not a very likely mechanism, either from a physical or from a physiological point of view. Furthermore, a number of empirical kinetic equations have been used, to which four of the seven parameters in the model have been fitted.

It is very interesting to compare three models. When the hyphal element continues to grow, a pellet can emerge. Yang et al. [53] and King [54]), working with *Streptomyces* – which is not a fungus but a member of the prokaryotes – and Bellgardt [6], working with *Penicillium,* start with the growth of hyphae but deal mainly with the growth of a pellet from a single spore. These models show a considerable amount of structure; they contain submodels for growth, septation and branching of hyphae, and simulate the observed phenomena very well. They show quite a few family features but clear differences as well.

The part of all these models that deals with hyphal growth is based on a model that combines diffusion with chemical reaction (DCR). The chemical reaction concerns the production – in one model the degradation as well – of a key precursor of growth. Production is either constant or contains a saturation term. Degradation is first order. For septation, a Trinci-like approach was used. The models contain several stochastic elements: branching occurs near the septa but is normally distributed, and the angles for branching and growth are normally distributed as well.

DCR models have also been used to calculate oxygen and other substrate profiles in the pellet. King [54] uses an purely kinetic equation for the elongation and fragmentation of hyphae and a DCR equation for the formation and fragmentation of tips in the pellet. In the model of Bellgardt [6], mycelial growth in the pellet is described the same way as hyphal growth. The diffusion coefficient in the DCR model for the substrate depends on the local density of biomass, and therefore on its radial position in the pellet. This model contains a term for the fragmentation of tips that grow out of the dense part of the pellet. This fragmentation was described with a probability function. Fragmentation of the hyphae occurs only at the tips.

Not too many parameters are involved in this modelling, and a number of them can be measured separately a priori.

The assumptions underlying these models are shown in Table 3. Table 3 shows how a limited number of basic mechanisms can result in rather differently structured models. The amount of structuring can still be increased, but at a cost as will be discussed in Sect. 4.2.5.

Table 3. Comparison of a number of highly structured models

	King	Bellgardt	Yang et al.
Hyphae			
Hyphal elongation eq.	DCR[a]	DCR	DCR
Hyphal elongation kinetics	Zero-order production 1st order degradation of key component	Zero-order production	Logistic production[b]
Transport in hypha	Passive diffusion	Passive diffusion	Passive diffusion
DNA replic.	Logistic production	n. i.	Logistic production
Septum formation	Deterministic	n. i.	Deterministic
Branching	Stochastic	Stochastic	Stochastic
Pellets			
Growth	From one spore	From one spore	From one spore
Hyphal elongation eq.	Kinetics only (no mass transfer)	See hyphae (above)	See hyphae (above)
Kind of kinetics	growth, degradation, fragmentation	See hyphae (above)	See hyphae (above)
Tip formation	DCR	Stochastic	Stochastic
Tip formation kinetics	Tip formation degradation, fragment	n. r.	n. r.
Substrate consumption	DCR	DCR	n. i.
Density	Growth dependent	$f(r_{crit})$	n. i.
Diff. Coef.	Based on void volume	f (cell density)	n. r
Fragmentation of tips	Included in kinetics	Probability function	n. i.

n. i., not included; n. r., not relevant.

[a] DCR: an equation containing terms for *d*iffusion and (*c*hemical) *r*eaction kinetics.

[b] Logistic equations are of the form: $r(C) = k.(C_{max} - C)$. There is also another definition of a logistic equation! [55].

As has already been mentioned, experimental confirmation of the models is not bad at all. This does not mean, however, that the models are correct from a physiological, or even physical, point of view (see Sect. 4.2.5). It is clear from Sect 4.2.1 that classical diffusion makes no sense. Yang [53] assumes explicitly that there is no active transport of the key component in the hyphae and zero concentration of the key component for hyphal elongation at the tip. This can only be true if the vesicles do not contain the key components. Furthermore, it is obvious that the kind of kinetics used for hyphal extension describes what is going on quite well, but its choice seems rather arbitrary. Interestingly enough, it does not seem to be important whether or not the production term for hyphal elongation is zero order or logistic, and whether a first order degradation term has been included. This raises an interesting, not merely philosophical, question: What do we mean when we say that a model is mechanistic?

To conclude this section, an impressive quantity of work has been published about the influence of environmental factors ($\mathbf{z}(t)$) on morphology. These results are mainly of an empirical nature, although suggested mechanisms are not uncommon. However, structured mechanistic models in this area, based on intracellular reactions at a molecular level, are virtually absent. No examples are presented here.

4.2.4
Models Describing the Relation Between Morphology and Production

Fungi are important workhorses in the fermentation industry, and models that allow for optimization of production with fungi can be very useful. In a special group of models morphology has been related to production. If the production of a component is dependent on the morphology of a microorganism as such, rather than via the influence of morphology on bulk transport, the use of morphologically structured models is a must. A number of models have been set up to describe this relationship, with the final goal of optimizing productivity. These models generally show a high degree of structure, and describe the production process quite well, at least at laboratory scale.

Three examples will be presented: first, an old one [4], followed by two recent ones [5, 6].

The first model [4] contains five "differentiation states", four of which are connected with a particular product, whereas the fifth is the active growing tip. The concepts of the model are suggested by data from experiments of the authors with *Aspergillus awamori*. The different morphological forms have been observed on surface cultures. A highly structured model was set up for batch and continuous cultures. The 43 parameters were obtained from the literature and their own work. The phenomena observed during the simulation find a number of parallels in the literature, but there is no direct comparison between their simulations and detailed data from parallel experiments. An extensive overview of this model has been given by Nielsen [56].

In a morphologically structured model for production of cephalosporin with *Streptomyces* on complex substrates [6], four different morphologies were discerned:

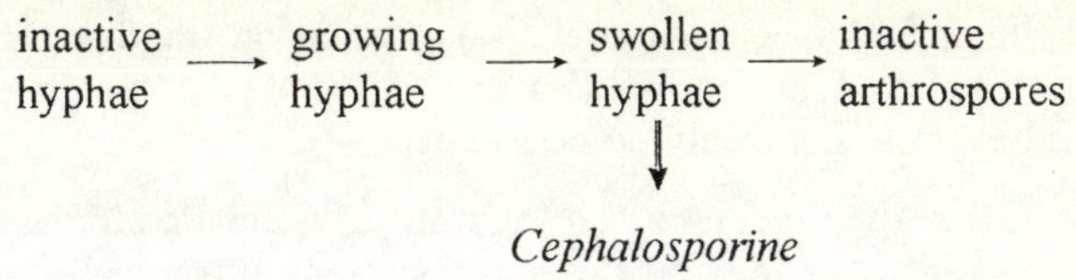

Fig. 5. The four morphological forms of *Acremonium chrysogenum*

The model consists of balances among these morphological forms, and a number of kinetic equations for growth and product formation and for substrate consumption with Michaelis–Menten-like substrate dependence. Repression of oil consumption in the presence of sugar is also included.

The model reproduces quite well the experimental time dependence of the concentrations of cell mass, oil and sugar substrate, product, and CO_2 in the off gas, also for several cultures. The model contains 20 parameters. It is not clear from this publication how these parameters were obtained. The model is applied to the dynamic optimization of feeding strategies.

If there are no morphological differences between hyphal elements, it is still possible to look at regions in hyphae with different activities [5]. Five different regions are distinguished in hyphae of *Penicillium chrysogenum*, based on different structures and activities. The model contains 20 parameters. Eleven parameters were obtained from the literature, 9 parameters were obtained from a continuous reference process. The fit between the simulation with the model and the data of this experiment was very good, but that is not surprising with nine parameters to tune the model to the experiments. However, the same model, with the original parameters obtained from the continuous culture, was used to simulate the outcome of a fed-batch culture. Again, the results were very good. This increases the value of the model. The sensitivity of the model to the various parameters used was also checked by means of a very simple – but very useful – technique for finding out which parameters are really important and need to be known with high accuracy and for which parameters evaluation to within an order of magnitude is sufficient. This technique has also been used by others [12].

4.2.5
Some General Remarks About Models

This is the place to make a few remarks about the models we saw and about models in general. A model is always a simplification of reality. It is, as someone once said, the art of balancing between unwieldy complexity against oversimplification. First, one quotation to "set the scene":

It should be noted, however, that a simulation giving realistic pictures does not necessarily prove the correctness of the assumptions used [53].

Almost 20 years earlier Topiwala [57] made a similar remark. In fact, models need be neither mechanistic nor structured in order to give a good correlation between model and experiment. Whether or not we have to worry about this depends on the objective of the model, in particular on whether this objective requires some form of extrapolation from the model, although even interpolation can be a risk. Important objectives are:

- Process control: Any model – including non-mechanistic models like neural networks – will do, as long as the process is not extrapolated beyond the range of conditions in which the model has been tested.
- Process optimization: The objectives are the same as for process control, but unreliable if the optimum lies outside the test range, i.e. requires extrapolation.
- Scale-up: Scale-up is extrapolation *par excellence*, and is therefore very unwise.
- Satisfaction of curiosity: Here, the goal is to unravel the real mechanisms, in order to construct a model that is solidly rooted in physiology and physics. Jeopardizing the model, by extrapolation, inter alia, is the essence of the game ; or "We should ring the bells of victory every time a theory has been refuted" (Popper [58]).

The objective determines a number of other issues that call for explicit decisions.

Time Dependence vs (Quasi) Steady State. This is a rather obvious decision to be taken. Time dependence can result in rather awkward differential equations, but very often a quasi-steady state approach yields acceptable results. For example, the concentration profile of O_2 in a growing pellet can be calculated very accurately without taking into account its momentary growth.

Number of Compartments. A compartment can be a geometrically defined compartment, an event, e.g. hyphal tip extension or septation, or a number of morphological forms – each with substrate consumption, product formation etc.

Mechanisms per Compartment. These mechanisms include the following:

- Transport: flow, diffusion (ordinary or active), pushing aside.
- Kinetics: nth order, logistic, saturation (e.g. Michaelis–Menten etc.).
- Equilibrium: thermodynamic, mechanical, saturation (full is full).

The choice of the mechanisms involved in each compartment is often rather arbitrary and not well argued. As stated above, this is not necessarily a problem. However, if the model has to be used for extrapolation or if the driving force is curiosity, full attention must be paid to this aspect. Some other aspects of the choice of mechanisms are as follows.

A number of mechanisms can be chosen, among others diffusion and flow, for the transport of vesicles to the tip of hyphae. These two mechanisms are compared in the Appendix. At first sight, both mechanisms give equally satisfactory results. Both are impossible, however. In order to see why, one has to look at ($C_{vesicle}(x)$) in greater detail.

Equations (8) and (10) in the Appendix can also be written as $C_1 = c_1 (L^2 - x^2)$ and $C_2 = c_2 x$, respectively, where c_1 and c_2 are constants. This "lumping of parameters" facilitates parameter optimization, but makes it difficult to jeopardize the model, because one loses sight of the constituent mechanisms.

Considering the enormous concentration of vesicles at the very tip of the apex, one might consider the possibility that something other than transport of vesicles, perhaps the inclusion of vesicles in the membrane, is rate-determining.

A rather extensive set of physiological events contribute to hyphal tip extension (see Sect. 4.2.1). In most publications, this boils down to a simple Michaelis–Menten kinetic equation for one component. Maybe this is correct, but it is seldom well argued.

As mentioned above, a good fit between model and experiment may be quite independent of the submechanisms, but it usually has a great deal to do with the next point.

Parameter Estimation. For extensively structured models, parameter estimation is often more time-consuming than the setup of the model itself. The problem is the number of parameters required by the model: roughly five parameters are needed per "compartment" If the compartment is geometrically defined, there are as many as three geometric parameters, depending on its shape; the diffusion coefficient and/or linear velocity; and at least two kinetic parameters. Thus, a model with four well-defined compartments needs 20 parameters [5] and one with three phenomena, e.g. hyphal extension, septation, and branching, needs 13, including those required for specifying initial conditions [53]. The important question is: "Where do these parameters come from?" Are they from "first principles", from the literature, independent measurements carried out separately in vitro and in vivo? Most important of all: How many parameters have been used to "fit" the model to experiment?

First, modelling the behaviour of a system and using the behaviour of the same system to determine the parameters is a form of inbreeding. As mentioned in the section on the choice of mechanisms, this is no problem if the model is to be applied in the same range of conditions as was used for determining the parameters.

If use of the model includes extrapolation, the situation is different. One should then jeopardize the model by working with the fungus under other environmental conditions ($\mathbf{z}(t)$), other process modes, e.g. batch, fed-batch or continuous, and/or other dilution rates or stirrer speeds. The agreement between the experimental data and the predictions of the model,without further adaptation of the parameters, is a good measure of its robustness.

Second, any system can be "modelled" with 20 or 30 parameters. An old, but still very relevant, publication in this respect has the intriguing title, "The least squares fitting of an elephant" [59].

The almost exponentially increasing performance/price ratio of computer systems makes parameter fitting easier and easier. Unfortunately, this is often at the expense of observations, because the performance/price ratio of modern analytical apparatus remains much more constant.

An often neglected problem is the effect of scale-up on the values of parameters. The morphology is very dependent on environmental conditions ($\mathbf{z}(t)$), which are themselves very often dependent on scale. In particular, a change of regime, for example, from kinetic to transport control, can be very misleading. The vast majority of parameters have been obtained from small-scale experiments, because production scale parameters are not usually available for publication.

The substrates on the laboratory scale and on the production scale often are – and have to be – quite different [60]. This also affects **z**(t).

The number of inoculation steps increases from 1 on the laboratory scale to about 4 on production scale. This too can influence the behaviour of the fungus.

These are typical scale-down problems: how to create a small-scale environment that is representative of the full-scale conditions.

In conclusion, the physiological basis of many models, including those in present use, has to a large extent been one publication [11], together with a small number of related publications. This publication has been exploited extensively, and for very good reasons, but we now need to make further advances. Today, tools are available for elucidation of the biochemical and genetic background of the mechanisms involved, although the task will not be an easy one. Close collaboration between those who construct models of the types mentioned above and mycological physiologists and geneticists would be a great help. A serious problem is that mycological physiologists are relatively scarce. In the Netherlands, for example, fewer than one in ten microbial physiologists is a mycologist and the ratio is probably not much better elsewhere.

4.3 Special Aspects

So far only methods and models have been presented that primarily allow for a better description and understanding of the morphogenesis of filamentous fungi. They also give some clues for the improvement of morphology. However, there are more methods for the improvement of the morphology (see Table 4 for an impression of the methods that could be used).

Some important issues can be extracted from Table 4 about how morphology can be influenced, but we will deal exclusively with genetics, because at present the other methods are obtained almost exclusively from measurements based on empirical know-how.

Furthermore, insight into the relation between morphology and transport phenomena in a mycelial reaction mixture can be of some help for the solution of morphology-based problems in production processes.

Table 4. Methods of influencing morphology

Genetics (either r-DNA or CSI)	Other methods
1. Primary cell wall synthesis	1. Inocculation
2. Altered metabolism	2. z (t) (including shear rate)

CSI is classical strain improvement.

4.3.1 *Genetics*

Morphology is a phenotypic property. The development of a phenotype, e.g., a fungus with a specific morphology, is always the result of the genotype and the environment, a relationship that can be formulated: Phenotype = Genotype × Environment, i.e. P = G × E. As mentioned in Sect. 4.2.3, there is an impressive quantity of literature, albeit fragmented, about the influence of environmental factors on morphology; as has been mentioned above, it is mainly of an empirical nature. However, very little knowledge, even empirical knowledge about the effect of genes on morphology is available. This is a real stumbling block, for if we want to master morphology we cannot do without genetics.

The genetics of morphogenesis can be very complicated. A non-fungus example of the effect of genes on morphology is the endospore formation of a "simple" bacterium: *Bacillus subtilis* [61]. At least 45 separate loci involved in endospore formation have been identified. Some consist of single genes, others consist of gene clusters. Also, the discussion of genetics in the introduction to physiology (Sect. 4.2) shows the complexity of the genetics and its control.

Morphology is determined to a large extent by the structure of the rigid cell wall. This means that mutations that affect this structure also affect morphology. However, other processes in he cell can also affect morphology and are candidates for consideration as factors that can be influenced by genetics. Although there is little hope for the elucidation of the detailed genetics behind morphogenesis of fungi in the near future, there are several leads to possible shortcuts. For example, changing a single gene can have a drastic effect on morphology.

1. Brody and Tatum [63] demonstrated that a point mutation in one gene, resulting in a change of affinity of a single enzyme (glucose-6-P dehydrogenase) to its substrate, effected a drastic change in the morphology of a surface culture of *Neurospora crassa,* probably because of in vivo accumulation of glucose-6-P. Furthermore, the change in the affinity of the enzyme for its substrate was caused by the alteration of a single amino acid in its structure. It is also interesting that changes in the source of carbon now had no obvious effect on the newly produced morphology. It has also been mentioned that the morphology of *Neurospora crassa* can be changed by altering the genes effecting cell wall synthesis [28].
2. Nonaka [62] mentions an interesting example of the effect of small genetic changes on the morphology of *Saccharomyces cerevisiae.* He investigated the effect of the protein Rho1p on the budding of *Saccharomyces cerevisiae.* Rho1p switches glucan synthesis on at budding and off at maturation of the daughter cell. Rho1-depleted mutants stop growing with small budded cells.

The experimental results with *Neurospora crassa* [63] and with *Saccharomyces cerevisiae* [62] show that changing one gene can have a dramatic effect on morphology. If these methods could be applied in some way to other fungi as well, the use of enrichment cultures would be an interesting option. This method is, of course, almost 100% empirical, but it may lead to a rapid improvement in the morphology of strains, and it may help find out what molecular biologists have to look for, and where. With some creativity, it should not be too difficult to set up a system that selects for the preferred morphology. This selection may take some time, but it may well pay off. The selection of a special form of microorganism (pellets), necessary for the large-scale anaerobic treatment of wastewater, took more than a year [64].

The overall effect must be a stable increase in productivity in the full-scale reactor, as a result of the combined positive effects of the genetics on the inherent productivity of the fungus as such, as well as on the transport properties, due to its effect on the improved morphology.

4.3.2
Whole Broth Properties

As seen in Fig. 3, there is an intricate relationship between viscosity and transport properties. Furthermore, it is well known that morphology has a strong influence on viscosity [9, 12]. Because poor transport properties can ruin full-scale productivity, these relations are very important (see Fig. 6, which is a variation on the theme of Fig. 2).

It appears to be difficult to represent the morphology of fungi unequivocally by rheological equations, although many attempts have been made by Metz [9, 12] and Olsvik [65] and their respective co-workers, among others. The next difficult step is finding the relation between the often rather peculiar viscosity equations and transport properties. The last step, going from transport properties to productivity, is an application of basic chemical engineering. This is route A in Fig. 6.

One can also save a lot of trouble and choose the more phenomenological approach of route B in Fig. 6, which calls for finding a direct relation between morphology and transport properties.

The reason for the difficulties of route A are threefold. First, the consistency – a better word than viscosity – of the filamentous mixture originates

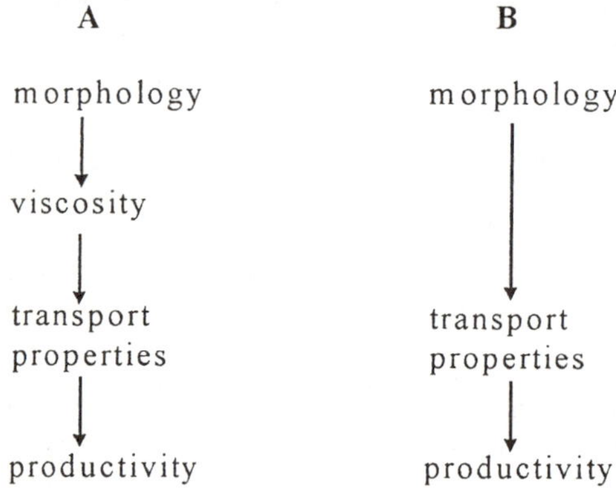

Fig. 6. From morphology to productivity

almost entirely from the interaction between solid particles. Second, these particles have almost macro-dimensions: from 0.1 mm for hyphae to several mm for flocks and pellets. Third, the consistency appears to be time-dependent.

This suggests that a dynamic corpuscular approach would be more successful than a static continuum approach. The latter is what one usually finds in the literature.

Some first steps have been made along route B [9]. Dynamic flocculation/deflocculation processes appear to be taking place in stirred mould suspensions. Deflocculation proceeds very rapidly under high-shear conditions, but in quiescent regions complete flocculation also occurs within about 0.5 s. This process time is only slightly dependent on dry mass-concentrations (C_X) in the range between 13 and 20 kg m^{-3}, but increases to $\cong$5 min for $C_X < 0.1$ kg m^{-3}. Although the flocculation/deflocculation process is very fast, a number of interesting phenomena can be observed in bubble columns.

When bubbles rise in a flat plate bubble column, the wake of a bubble has an appearance that is different from the bulk of the broth. This effect may be the result of demixing alone or of demixing with its subsequent influence, due to the lower local value of X on the breaking and reforming of the structure of the broth. In any case, if successive bubbles are released, one can see that the first bubble travels much more slowly than the second. This second bubble obviously profits from the path made by the first, and so on. The overall effect is a very rapid coalescence of bubbles, growing from 5 mm to 5 cm in diameter within a 1-m column. In some full-scale mycelial processes, with a very interactive mycelium at high biomass concentrations, this results in the aforementioned applesauce appearance of the reaction mixture and football-sized bubbles, bubbles that would be too unstable to ever exist in a homogeneous fluid. Mass transfer would be extremely slow if it were completely dependent on these bubbles.

Mass transfer in a thick mycelial reaction mixture is in practice not as bad a mass transfer mechanism as might be expected. It has therefore been suggested [9] that mass transfer is not based on the contribution of bubbles, but on the intense breakup and oxygen saturation of flocs near the stirrer. The resulting equation has the following form:

$$k_1 a = \frac{\Phi_p}{V} = \frac{1}{t_c} \cong \frac{N \cdot D^3}{V} \tag{5}$$

and gives the right order of magnitude for k_1a in full-scale reactions.

Another aspect of the structure of the broth is the entrapment of small bubbles in the mycelial network. This acts as a sink or source for O_2 and CO_2.

This kind of modelling has obviously not generated many followers.

Problems with the measurement of viscosity of filamentous reaction mixtures are also the result of the inhomogeneous character, as was mentioned above.

The measurement of rheology has improved since the first measurements with a turbine impeller in the laminar flow region [9, 12, 66]. Experiments with classical rotating cylinders for rheology measurements were unsatisfactory for three reasons:

1. The particles are of the same order of magnitude as the annulus, resulting in damage of pellets and flocs during measurement.
2. Less dense layers are formed near the wall.
3. The particles have a tendency to settle.

Tube viscometers avoid problem 1, but need a relatively large diameter when pellets as large as 1 mm are used, in which case large samples are needed [66].

Allan and Robinson [67] have shown, with suspensions of three different moulds, that a helical ribbon is preferable to the turbine impeller for viscosity measurements, because the assumption that the average shear rate in the turbine viscometer is independent of fluid rheology is not correct – at least not for fluids with a low flow-behaviour index (n) in the power-law equation:

$$\tau = K \cdot \dot{\gamma}^n \qquad (6)$$

and for fluids with a yield stress. Their rotating cylinder viscometers and pipes with $D > 10$ mm worked well, contrary to earlier experience [66]. This could be because different strains and suspensions without pellets were used [67].

An interesting issue is the relation between the rheological properties and the morphology of the hyphal clumps. A positive correlation was found between the roughness of these clumps, i.e. aggregates of micelium, and the biomass concentration of *Aspergillus niger*, on the one hand, and the consistency index K of the power law equation, on the other [65]. This correlation holds for various values of the growth rate and the dissolved oxygen tension (DOT) and for two different values of the biomass concentration.

5 Implementation of the Results

The amount of research carried out in the past on the morphology of filamentous fungi has been impressive. However, the implementation of these results in the fermentation industry seems to be limited. There are a number of reasons for this limited implementation:

The first reason is the problem of time squeeze. In the fermentation industry, one is constantly looking for more productive strains. Per product, this results in about one new, more productive strain every few years. This strain is taken into production as soon as possible, and any morphology problems are solved by empirical methods based on extensive experience. Schügerl [20] points out the complex interrelationships between all the factors involved in growth, morphology, physiology and productivity of moulds, and states: "Without considering all of the relevant parameters it is not possible to make general conclusions". That is correct, but unfortunately, under the pressure of economics, including time, new strains will have been developed long before the measurement of all of the relevant parameters are complete.

The second reason is a problem of genetics. The genetics of the production of an enzyme in microorganisms, including fungi, is far more developed than the genetics of morphology where many enzymes, and their corresponding genes, are involved (see the introduction to Sects. 4 and 4.3.1). What is wanted is a kind of Holy Grail: a mould with the morphology of yeast and without loss of inherent production capacity.

The third reason is a screening problem. Mass screening for a better productivity is relatively easy, there are many assays and hundreds of thousands of new, mutated strains can be produced and tested per year for any given product. However, no assays that perform mass screening for a better morphology can handle so many mutant strains.

6
Conclusions and Prospects

1. Powerful new research tools have been developed, resulting in greater insight into the morphological details of fungi and an increase in the quantity of information and its quality.
2. Highly structured models exist for morphogenesis and for the relation between morphology and productivity. The application of these models in industry seems to be rather limited for a number of practical reasons.
3. The mechanistic postulates underlying these models vary from some that are firmly based on sound physiological principles to others that are in conflict with these principles.
4. Whether or not it is important that a model have a sound mechanistic background depends on the purpose of the model and whether or not it will be used for extrapolation.
5. The physiology of the morphogenesis of fungi is making progress, but the know-how of the genetics behind it is very limited. This is most unfortunate, because a sound genetic base is very important for the future development of fungi with a high inherent productivity combined with morphological properties that result in high rates of momentum, mass, and heat transport.
6. Mass screening and enrichment cultures for favourable morphologies may possibly fill the gap until the basic genetics for morphology has been developed sufficiently far.
7. The control of morphology should be based on insight into genetics, physiology and biochemical engineering ,and on real integration of those three areas, in particular among the people working in each of these areas on one project.
8. Other methods of overcoming the morphology problems might be:
   a. Growing fungi on carrier particles in the form of pellets or as layers. This has been the subject of many studies, but here too the application in industry is limited.
   b. The use of other microorganisms, such as yeasts or bacteria.
   c. The use of moulds in solid state processes.
   d. The use of plants [68, 69]. This point needs some explanation. Production of microbial enzymes, such as phytase, in plants has been proven to be feasible. Phytase is a very interesting enzyme for the manufacture of fodder for pigs and poultry, because it reduces the phosphate content of the

manure considerably. The enzyme was originally produced in *Aspergillus*. The expression of the phytase gene in plants can be made tissue-specific. Its expression in seeds results in a product with relatively high enzyme concentrations. The product is very stable, free-flowing and non-dusting This method has a broad and very interesting potential for applications.
9. To end with a special remark. The school of Trinci has been standing like a beacon in the landscape of morphology of fungi for a number of decades.

Acknowledgements. The author wishes to thank Dr. Sietsma, of the University of Groningen, for his positive criticism and his additions to the introduction of the physiology of the growth of fungi, and Dr. Krabben, presently a post-doctoral student at the Delft University of Technology, for numerous of discussions. The author remains fully responsible for any remaining faux pas.

Appendix

To show how little influence the kind of model can have on the outcome of a simulation, two very simple models for the growth of hyphae due to the transport of vesicles to the tip are compared. In one model, transport is based on diffusion; in the other model, transport is based on flow.

Diffusion Model. The assumptions are: transport of vesicles takes place by ordinary diffusion, and the vesicles are formed with 0th order kinetics. The rate of formation is given by Prosser and Trinci [11]: $r_l = 1.5$ vesicles μm^{-1} min^{-1}. The value of r_l is not important; the important point is that r_l is constant. The hyphal element consists of one hypha. In the middle of the hypha the concentration gradient of the vesicles is zero; at its tip their concentration is zero, due to rapid uptake of vesicles by the wall of the apex.

The assumptions lead to the following differential equation and boundary conditions:

$$-ID \cdot \frac{d^2C}{dx^2} = \frac{4}{\pi \cdot d_h^2} \cdot r_l \tag{7}$$

The boundary conditions are $dCx/dx = 0$ for $x = 0$ and $C = 0$ for $x = L$

The solution of this equation is:

$$C = \frac{2 \cdot r_l}{\pi \cdot d_h^2 \cdot ID} \tag{8}$$

Flow Model. The assumptions are: transport of vesicles takes place by constant flow. The rate of formation of vesicles is identical to that in the diffusion model. In the middle of the hypha the concentration is zero.

These assumptions lead to the following differential equation and boundary condition:

$$v \cdot \frac{dC}{dx} = \frac{4}{\pi \cdot d_h^2} \cdot r_l \tag{9}$$

The boundary condition is $C = 0$ for $x = 0$.

The solution is:

$$C = \frac{4 \cdot r_l}{\pi \cdot d_h^2 \cdot v} \tag{10}$$

For both equations, the flow at $x = \mathrm{L}$ is equal to $r_l \cdot \mathrm{L}$ That is as it should be, because this is the total amount of vesicles formed per second over the length L, and – in the steady state – it should be equal to the amount arriving at $x = \mathrm{L}$. Because the rate of growth of the hypha is proportional to $r_l\mathrm{L}$, both models predict exactly the same rate of growth. Needless to say, this is independent of the value of *ID* or *v*!

These models can also be used to calculate the apparent velocity due to diffusion of the vesicles. If one calculates the diffusion coefficient *ID* with the Einstein equation and assumes that the observed transport velocity for a diffusion process can be calculated with

$$v_{observed} \cdot C = -ID \frac{dC}{dx} \tag{11}$$

where C and dC/dx can be calculated from Eq. (1), then the observed velocity is of the same order of magnitude as that found by Prosser and Trinci [11], about 10 µm min^{-1}.

It is clear that we cannot conclude which model is the correct one from a mechanistic point of view by comparing the simulated rates of growth or the transport velocities with the experimental values. However, if we look at the concentration gradients of vesicles, as observed by Collinge and Trinci [70], then the picture changes completely. In fact, the values of $C(x)$ calculated

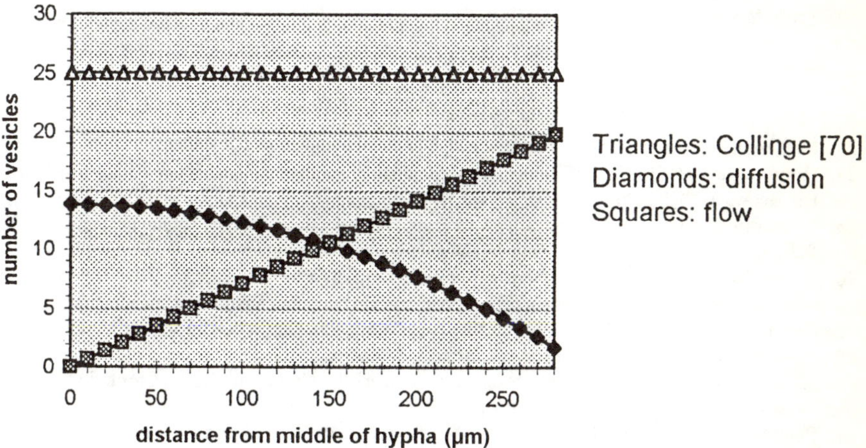

Fig. 7. Concentration of vesicles (number/µm^3) as a function of the distance to the middle of the hypha in µm.

by both models conflict with the measurements of $C(x)$ for *Neurospora crassa* where $C(x)$ is almost constant at about 25 vesicles per μm^3 until the apical compartment has been reached, where it increases sharply.

Both models presented above are gross oversimplifications of reality and are therefore not very realistic. However, the message is clear: without close observation of reality, the models for the growth of hyphae cannot be considered to be mechanistic in the physical/physiological sense.

References

1. Scheper T, Schügerl K (eds) (1998) Adv Biochem Biotechnol, vol 60, Springer, Berlin Heidelberg New York
2. Nielsen J, Villadsen J (1994) Bioreaction engineering principles. Plenum Press, New York
3. Metz B, Kossen NWF, van Suijdam JC (1979) Adv Biochem Eng Biotechnol 1:103
4. Megee RD, Kinorhita S, Fredrickson AG (1970) Biotechnol Bioeng 12:771
5. Paul GC, Thomas CR (1996) Biotechnol Bioeng 51:558
6. Bellgardt KH (1998) Process models for production of β-lactam antibiotics. In: Scheper T, Schügerl K (eds) Adv Biochem Biotechnol 60:153
7. Dion WM, Carilli A, Sermonti G, Chain EB (1954) Rend Ist Super Sanita 17:187
8. Dion WM, Kaushal R (1959) Sel Sci Pap Ist Super Sanita 2:357
9. van Suijdam JC, Metz B (1981) Biotechnol Bioeng 23:111
10. Fiddi C Trinci APJ (1976) J Gen Microbiol 97:169
11. Prosser JI, Trinci APJ (1979) J Gen Microbiol 111:153
12. Metz B (1976) From pulp to pellet. Dissertation, Delft University of Technology
13. Metz B, de Bruin EW, van Suijdam JC (1981) Biotechnol Bioeng 23:149
14. Adams HL, Thomas CR (1987) Biotechnol Bioeng 32:707
15. Paul GC, Thomas CR (1998) Characterisation of mycelial morphology using image analysis. In: Scheper T, Schügerl K (eds) Adv Biochem Biotechnol 60:1
16. Paul GC, Kent CA, Thomas CR (1992) Trans I Chem Eng (Part C) 70:13
17. Pons MN, Vivier H (1998) Beyond filamentous species. In: Scheper T, Schügerl K (eds) Adv Biochem Biotechnol 60:61
18. Krabben P, Nielsen J (1998) Modelling the mycelium morphology of penicillium species in submerged cultures. In: Scheper T, Schügerl K (eds) Adv Biochem Biotechnol 60:125
19. Spohr A, Dam-Mikkelsen C, Carlsen M, Nielsen J, Villadsen J (1998) Biotechnol Bioeng 58:541
20. Schügerl K, Gerlach SR, Siedenberg D (1998) Adv Biochem Biotechnol 60:195
21. Nielsen J, Johansen CL, Villadsen J (1994) J Biotechnol 38:51
22. Thiele EW (1939) Ind Eng Chem 31:916
23. Rvesbech NP, Jørgensen BB, Brix O (1981) Limnol Oceanogr 26(4):717
24. Hooijmans CM (1990) Diffusion coupled with bioconversion in immobilized systems. Dissertation, Delft University of Technology
25. Cronenberg CCH, Ottengraf SPP, van den Heuvel IC, Pottel F, Sziele D, Schügerl K, Bellgardt KH (1994) Bioproc Eng 10:209
26. Cronenberg CCH (1994) Biochemical engineering on a micro-scale:biofilms investigated with needle type glucose sensors. Dissertation, University of Amsterdam
27. Gooday GW, Lloyd D, Trinci APJ (1980) 13th Symp Soc Gen Microbiol, p 207
28. Scott WA, Tatum EL (1970) Proc Nat Acad Sci USA 66:515
29. Schlegel HG (1993) General microbiology. Cambridge University Press, Cambridge, p 170
30. Howard RJ, Aist JR (1980) J Cell Biol 87:55
31. Regalado CM, Sleeman BD, Ritz K (1997) Philos Trans R Soc Lond 352:1963
32. Cabib E, Roberts R, Bowers B (1982) Annu Rev Biochem 51:763
33. Drgonová J, Drgon T, Tanaka K, Kollár R, Guang-Chao Chen, Ford RA, Chan CSM, Takai Y, Cabib E (1996) Science 272:277

34. Kamada Y, Qadota H, Python CP, Anraku Y, Ohya Y, Levin DE (1996) J Biol Chem 271:9193
35. Yamochi W,Tanaka K, Nonaka H, Maeda A, Musha T, Takai Y (1994) J Cell Biol 125:1077
36. Wessels JGH (1993) Advances in microbial physiology 34:147
37. Fredrickson AG, Tsuchia HM (1963) AIChE J 9:459
38. Randolph AD (1964) Can J Chem Eng 280
39. Randolph AD, Larson MA (1971) Theory of particulate processes. Academic Press, New York, p 41
40. Nielsen J (1993) Biotechn Bioeng 41:715
41. Ainsley M, Ward AC, Wright AR (1990) Biotechnol Bioeng 35:820
42. Bergter F (1978) Z Allg Mikrobiol 18:143
43. Trinci APJ (1970) Arch Microbiol 73:353
44. Trinci APJ [1970] Trans Br Mycol Soc 55:17
45. Plomley NJB (1959) Aust J Biol Sci 12:53
46. Caldwell IY, Trinci APJ (1973) Arch Microbiol 88:1
47. Emerson S (1950) J Bacteriol 60:221
48. Nielsen J, Krabben P (1995) Biotechnol Bioeng 46:588
49. Hinze JO (1975) Turbulence. McGraw Hill, New York, p 221
50. van Suijdam JC (1980) Mycelial pellet suspensions. Dissertation, Delft Univerity of Technology
51. van Suijdam JC, Metz B (1981) J Ferm Technol 59:329
52. Ayazi Shamlou P, Makagiansar HY, Ison HY, Lilly MD, Thomas CR (1994) Chem Eng Sci 49:2621
53. Yang H, King R, Reichl U, Gilles ED (1992) Biotechnol Bioeng 39:49
54. King R (1998) Mathematical modeling of the morphology of streptomyces species. In: Scheper T, Schügerl K (eds) Adv in Biochem Eng Biotechnol 60. Springer, Berlin Heidelberg New York, p 95
55. May PN (1974) Stability and complexity in model ecosystems. Princeton University Press, Princeton, NJ
56. Nielsen J (1992) In: Scheper T, Schügerl K (eds) Adv Biochem Eng Biotechnol 46. Springer, Berlin Heidelberg New York, p 187
57. Topiwala HH (1973) Methods Microbiol 8:35
58. Popper KR (1946) Lecture, Signific. Congress Bussum, Holland (unpublished)
59. Wei J (1975) Chemtech Feb: 128
60. Kossen NWF (1993) Scale-up strategy in fermentation. In: Mortensen U, Norman HJ (eds) Proceedings of the International Symposium on Bioreactor Performance, Helsingør
61. Brock TD, Madigan MT (1991) Biology of microorganisms. Prentice Hall, Englewood Cliffs, NJ, p 18
62. Nonaka H, Tanaka K, Hirano H, Fujiwara T, Kohno H, Umikawa M, Mino A, Takai Y (1995) EMBO J 4:5931
63. Brody S, Tatum EL (1966) Proc Nat Acad Sci USA 290
64. Lettinga G (1973) Agricultural University Wageningen, personal communication
65. Olsvik E, Tucker KG, Thomas CR, Kristiansen B (1993) Biotechnol Bioeng 42:1046
66. Bongenaar JJTM, Kossen NWF, Metz B, Meijboom FW (1973) Biotechnol Bioeng 15:201
67. Allen DG, Robinson CW (1990) Chem Eng Sci 45:37
68. van Ooijen AJJ, Rietveld K, Hoekema A, Pen J, Sijmons PC, Teunis C, Verwoerd TC, Quax WJ (1996) US patent 5, 543, 576
69. van Ooijen AJJ, Rietveld K, Hoekema A, Pen J, Sijmons PC, Teunis C, Verwoerd TC 1997) US patent 5, 593, 963
70. Collinge AJ, Trinci APJ (1974) Arch Microbiol 99:353

Received December 1998

Antibiotica Research in Jena from Penicillin and Nourseothricin to Interferon

Harald Bocker, Wolfgang A. Knorre

Hans-Knöll-Institute for Natural Products Research, Beutenbergstraße 11, 07745 Jena, Germany
Fax: +49 3641 656800

Milestones of antibiotics research and biotechnology in Jena/Thuringia are: 1938 – Hans Knöll established a strain collection of microorganisms; 1942 – production of penicillin on laboratory scale by Hans Knöll; since 1945 – development of industrial production processes for penicillin and streptomycin; 1952 – production of BCG-vaccine; since 1956 – development of biotechnical processes in the Institute of Microbiology and Experimental Therapy for actinomycin C, oxytetracyclin, erythromycin, paromomycin, turimycin, griseofulvin, nystatin, and nourseothricin, and in the 1980s for streptokinase, staphylokinase, and interferons. After the German unification the Hans-Knöll-Institute for Natural Products Research was founded.

Keywords. Bioprocess development, Penicillin, Streptomycin, BCG-vaccine, Nourseothricin, Lysin, Streptokinase, Staphylokinase, Interferons.

In 1937 the well-known glass factory Jenaer Glaswerk Schott & Gen. started cooperation with the young physician Dr. Hans Knöll, living in Frankfurt (Main), in order to check their all-glass bacterial filters. These filters were produced for the first time in 1935 according to an invention of Dr. Paul Prausnitz, Head of the Department for Design and Manufacture of Apparatus, in this glassworks. Knöll had already dealt with the problems of tuberculosis and chemotherapeutics. He subsequently became known for the filtration of bacteria. On behalf of this company, Knöll developed an accurate measuring procedure for checking such filters. Being very interested in this, the Schott-factory offered him the opportunity to establish and manage a bacteriological laboratory in the glassworks. Knöll started this job on 1st November 1938, not being aware at this time that his work would become of special importance both for him and for future biotechnology activities in Jena.

Knöll established a still existing collection of defined strains of different microorganisms as a basis for filter checking, and further works in the fields of microbiology, chemotherapy, and cell biology. His activities in identifying new methods in phase contrast and fluorescence microscopy led to cooperation with the precision-mechanical-optical factory Carl Zeiss in Jena.

Through reference to the literature, Knöll's attention was drawn to penicillin, which had been discovered by Fleming in 1928 and which had been isolated by a research team in Great Britain in 1939. He started experiments to obtain this new antibiotic in mold cultures of *Penicillium notatum*. After a short time crude samples of this antibiotic were obtained. Originally Knöll intended to test penicillin for its effectiveness against cancer cells. However, the immense importance of penicillin in the fight against several human bacterial infectious

Advances in Biochemical Engineering/
Biotechnology, Vol. 70
Managing Editor: Th. Scheper

diseases soon became known. Penicillin-producing mold strains were sought in the environment and were cultivated in flat glass flasks such as Fernbach flasks. The use of Schott-borosilicatglass proved to be advantageous. It was found that the use of cheaper glass-types spoiled the synthesis of this antibiotic because of its arsenic content. Because of this, and in spite of wartime, an effective exchange – via foreign countries – of information about penicillin took place between Schott-glass and Jena.

Soon afterwards, penicillin wound powder was available from Jena on the laboratory scale. In late 1942, for the first time, it was applied to man. A factory worker, who had a suppurating injury to his hand, was cured successfully by application of this penicillin produced in the Bacteriological Laboratory.

In 1944, this successful work, together with support from the Carl Zeiss factory, led to the transformation of the Bacteriological Laboratory, whose staff had increased from 4 to 15 employees, into the Institute of Microbiology (Schott-Zeiss-Institute), supported by the founding firms of Schott and Zeiss.

At the end of the Second World War in June 1945, the US Army administration at that time in Jena intended to transfer the Institute of Microbiology to the western part of Germany. However, this intention was not realized as in July 1945 the occupation by the Soviet Army began. This new military administration ordered an immediate expansion in penicillin production. Cultivation of the producing mold had been intensified, while stage fermentors made from glass, aluminum, and steel, respectively, were installed in an empty factory building belonging to the firm Carl Zeiss. Railway tank wagons were also modified into fermentors. A great deal of effort was put in to erecting a production plant for penicillin attached to the Institute.

As a result of the rapid increase in the size of the operation, the fermentation section of the Institute of Microbiology was named Jenapharm in 1947. To improve the unsatisfactory supply of medicines, other medicaments and drugs, such as vitamins, analgesics, and transfusion solutions were incorporated into its production program. The number of employees grew to more than 800. Finally, in 1950 the Institute of Microbiology became an independent nationally owned factory, the VEB Jenapharm (VEB means a state-owned company). In addition to penicillin, the VEB Jenapharm produced another antibiotic, streptomycin, which was used to fight tuberculosis, incidence of this illness having increased considerably as a result of war.

A few months after the foundation of the German Democratic Republic (GDR), as a measure in the fight against tuberculosis, the Ministry of Health ordered Knöll to start the immediate production of vaccines according to the methods of Calmette and Guérin as a prerequisite for the introduction of BCG-vaccination. The first research building on the Beutenberg Hill in Jena was therefore erected as a production unit (see Fig. 1), where from 1952 onwards the BCG-vaccine was produced for the whole of the GDR.

In 1953, Knöll left VEB Jenapharm to become director of the newly founded Institute of Microbiology and Experimental Therapy (IMET), which was built in accordance with his ideas, also on the Beutenberg Hill. This institute was taken over by the German Academy of Sciences (later Academy of Sciences of

Fig. 1. Hans Knöll (1913–1988, founder of the Institute of Microbiology and Experimental Therapy on the Beutenberg and the pharmaceutical industry in Jena) left of Werner Eggenrath (Prime Minister of Thuringia) at the topping-out ceremony of the Microbiology Institute under construction in 1951. Knöll was awarded the National Prize of the German Democratic Republic. His Institute became a refuge for the politically displaced. Today the name of Hans Knöll is synonymous with Jena as are the names of Carl Zeiss and Otto Schott

the GDR) in 1956 and was transformed into the Central Institute of Microbiology and Experimental Therapy (ZIMET). Prof. Knöll was director of the ZIMET up to 1976. During this time the personnel at the institute increased to about 1000 workers.

The task of ZIMET was to work on therapeutics, particularly on microbial agents for use in human and veterinary medicine, and later it took over additional technical tasks. The structure of this institute incorporated all the research requirements necessary under the same roof in order to reach a high level of self-sufficiency. ZIMET was composed of the divisions Antibiotic Research, Biotechnology, Experimental Therapy, Medical Microbiology, Methods and Theory, Molecular Biology and Microbial Genetics, Steroid Research, Environmental Microbiology, and Scientific Engineering.

To maintain a continuous line of investigations from screening and increasing efficiency to testing the isolates and purified final products on animals, it was necessary to install qualified microbial and chemical laboratories as well as an efficient experimental breeding system, technical media preparing groups, and different workshops.

Antibiotics research in all its complexity and relevant applications became an essential task of this institute. In the 1950s further improvement in the production of penicillin and streptomycin, in cooperation with VEB Jenapharm, was its main objective. Later on the research potential was systematically developed, including corresponding fields of basic research.

For many years, both closely cooperating divisions "Antibiotic Research" and "Biotechnology" were mainly involved in the elaboration of specific bioprocess methods and down-stream processing by contract with the pharmaceutical industry of the GDR. Further shared research works were the search for producing microbial strains, and the developing of technical instructions for the biosynthetic production and chemical isolation of various antibiotics and other substances for therapeutic and technical purposes, respectively.

The behavior of microbial production strains in shaking flasks and laboratory fermentors was investigated to optimize process conditions and the composition of media on the basis of process kinetic analyses, as well as to elucidate the importance of certain medium compounds for special types of biosynthesis. For such investigations a new biometric screening method, based on the $(2n+1)$-spectrum, was developed and used with high efficiency.

In the 1960s basic investigations into growth and product-formation kinetics, as well as metabolic regulation in microorganisms, were started, aiming at scientific progress in the optimization of production methods obtained from models. On model systems, in a number of instances, the bistability of specific product formations of microbial processes was demonstrated successfully. Methods for optimal control of biotechnological processes were developed on the basis of mathematical descriptions of growth, metabolism, and product formation in microorganisms as well as by computer simulation of kinetic models.

In a pilot plant with reactors up to 3 m^3 net capacity the biotechnical methods were further adapted to the conditions of particular industrial production. Subsequently, the biosynthetically formed substances were obtained by selected known down-stream processes. Moreover, large numbers of small quantities of microbial agents for experimental purposes were produced there.

A widely recognized way to rationalize the multistage procedures of screening and selection, respectively, was opened by introducing six types of selection machine developed by an automation team of the ZIMET. This so-called Autoselect System includes machines to deliver small quantities of agar, to inoculate colonies, to dilute samples, to punch test plates, to pour sample solutions into the punched holes, and to measure the diameters of inhibition zones on the test plates by an optoelectronic method. Thus, by means of the Autoselect Systems, the number of colonies and samples tested could be considerably increased, and the accuracy of working steps and measurements was remarkably improved. The data were evaluated by computer. The optoelectronic measuring device was adopted by the production program of the factory VEB Carl Zeiss Jena.

The search for new antibiotics producing microorganisms had been a permanent task of the ZIMET, particularly of the Antibiotic Research division. Up to the 1980s a collection with more than 20,000 taxonomically identified, freeze-dried strains with defined antibiotic activity had been established.

New production methods were developed and transferred to industrial production for antibacterial antibiotics, such as actinomycin C, streptomycin, oxytetracyclin, erythromycin, paromomycin, and turimycin, as well as for the antifungal antibiotics griseofulvin and nystatin and the nutritive antibiotic nourseothricin, which was required in animal nutrition. Remarkable results, which were used in the industrial production of nourseothricin, were obtained from basic investigations on the regulating effect of phosphate on secondary metabolism. It was detected, that a regulated feeding with phosphate, liquefied starch, and ammonia, while realizing a phosphorus limitation and simultaneously sufficient concentrations of the other components in the medium, can increase the biosynthetic production of this antibiotic in stirred and intensively aerated bioreactors up to yields of 50 g/l nourseothricin. Furthermore, an economic mode of precipitating this agent, basing on its adsorption to bentonite, was invented and applied in industrial production.

Taking multivalent advantages of growth media, an example was set in the industrial streptomycin processes. By a specific change in down-stream processing, a proteolytic enzyme complex could be obtained in addition to the main product, streptomycin. This by-product, after concentrating in its aqueous solution, was used on the industrial scale in the VEB Filmkombinat ORWO Wolfen (Saxony-Anhalt) for the de-gelatination of photographic films in order to recover silver residues. Furthermore, the important VEB Lederfabrik Weida (Thuringia) used this product as a softening-enzyme in the leather-tanning process.

A new method of processing was adapted and tested successfully in the contamination-free production of L-lysine yielding an important food supplement in animal nutrition. In order to guarantee the required high oxygen transfer rate, deep-stream pilot reactors (0,45 m^3 capacity) were used. In cooperation with the Research Center of Biotechnology Berlin the developed processing was transferred to industrial application. In the VEB Gärungschemie Dessau (Saxon-Anhalt) non-contaminated production was realized constantly with yields of 80–90 g/l in a production time of 70 h, using a technical deep-stream bioreactor-device with a capacity of 8 m^3, fitted out with a continuously acting sterilization-unit for nutrient media. Such a high yield of L-lysine from a bioprocess without contamination was outstanding at that time.

In the early 1980s the ZIMET was concentrating its efforts on the further extension of biotechnology, particularly in the important field of the development of microbial agents from genetically engineered microorganisms. In this connection, genetic engineering protein technology and process development for rDNA products has been promoted.

The genes coding for the plasminogen activators streptokinase and staphylokinase were cloned and sequenced. Expression studies in a variety of hosts led to the construction expression vectors based on staphylokinase signals. Several bacterial strains producing significant amounts of human interferons alpha 1, alpha 2, and gamma have been constructed on the basis of these vectors. In order to maximize the production of interferons in the *E. coli* cells a high cell density bioprocess for the industrial application was developed in 1985. In 1987, after only two years, the first lots (about 10^{10} IU) of purified human inter-

Table 1. Institutes of the Academy of Sciences and of the Academy of Agriculture of the former GDR with R&D in Biotechnology

- Zentralinstitut für Mikrobiologie und experimentelle Therapie, Jena
- Zentralinstitut für Molekularbiologie, Berlin-Buch
- Institut für Biotechnologie, Leipzig
- Institut für Biochemie der Pflanzen, Halle
- Zentralinstitut für Genetik und Kulturpflanzenforschung, Gatersleben
- Zentralinstitut für Ernährung, Potsdam-Rehbrücke
- Zentrum für wiss. Gerätebau, Mytron, Heiligenstadt
- Forschungszentrum Dummersdorf bei Rostock
- Institut für Biotechnologie, Potsdam
- Institut für Getreideforschung, Hadmersleben
- Institut für Phytopathologie, Aschersleben

Table 2. Universities with scientific and technical activities in biotechnology

- Humboldt-Universität, Berlin
- Karl-Marx-Universität, Leipzig
- Martin-Luther-Universität, Halle
- Friedrich-Schiller-Universität, Jena
- Wilhelm-Pieck-Universität, Rostock
- Ernst-Moritz-Arndt-Universität, Rostock
- Technische Universität, Dresden
- Ingenieur-Hochschule Köthen
- Technische Hochschule, Leuna-Merseburg

feron alpha 1 were supplied to VEB Arzneimittelwerk Dresden for clinical trials phase I.

The Genetic Engineering Pilot Plant building, erected at the end of the 1980s, houses both a process development base for genetic technologies and a manufacturing unit. These provided researchers with capabilities for producing agents in compliance with applicable GMP regulations and for developing new process steps.

The biotechnological activities in the ZIMET were finished in 1991 by the "Abwicklung" of Academy of Sciences of the GDR. In 1992 the Hans-Knöll-Institute for Natural Products Research (HKI) was founded. It is one of five successor institutes of the ZIMET on the Campus Beutenberg in Jena/Thuringia. The other four successor institutes are Institute of Molecular Biotechnology, Institute of Molecular Biology, Institute of Experimental Microbiology, and Institute of Virology.

Biotechnological research and development in the GDR was concentrated in the Academy of Sciences, the Academy of Agriculture and in the Universities. Tables 1 and 2 give an overview of the Research Institutes and Universities of the GDR. According to the "Einigungsvertrag" of the German unification, all the institutes (Table 1) finished their work at the end of 1991.

Received January 2000

Development of Bioreaction Engineering

Karl Schügerl

Institute for Technical Chemistry, University of Hannover, Callinstrasse 3, D-30167 Hannover, Germany
E-mail: schuegerl@mbox.iftc.uni-hannover.de

In addition to summarizing the early investigations in bioreaction engineering, the present short review covers the development of the field in the last 50 years. A brief overview of the progress of the fundamentals is presented in the first part of this article and the key issues of bioreaction engineering are advanced in its second part.

Keywords. Fluid dynamics, Mass and energy balances, Process monitoring and control, Mathematical models, Metabolic engineering, Expert systems.

1 Introduction 43

2 Fundamentals 45

2.1 Fluid Dynamics and Transport Processes 46
2.2 Macroscopic Total Mass, Elemental Mass, Energy and Entropy Balances 48
2.3 Kinetics of Growth and Product Formation 48
2.4 Metabolic pathways 49
2.5 Process Monitoring and Control 49
2.5.1 pO_2 and pH Measurement 49
2.5.2 Biosensors 49
2.5.3 On-line Sampling, Preconditioning and Analysis 50
2.5.4 Process Control 51
2.6 Mathematical Models 51

3 Interrelation Between Physical, Chemical and Biological Processes 52

3.1 Influence of Fluid Dynamics and Transport Processes on Microbial Cultures 53
3.2 Process Identification by Advanced Monitoring and Control 57
3.3 Metabolic Engineering, Metabolic Flux Analysis 57
3.4 Expert Systems, Pattern Recognition 59

4 Particular Systems 60

4.1 Immobilized Microorganisms 60
4.2 High Density Cultures of Microorganisms 62
4.3 Animal and Plant Cell Cultures 62

5 Conclusions 64

References 64

Advances in Biochemical Engineering/
Biotechnology, Vol. 70
Managing Editor: Th. Scheper

List of Symbols and Abbreviations

a	specific interfacial area
$D_L(r)$	axial liquid-dispersion coefficient
$D_r(r)$	radial liquid-dispersion profiles
$d_{Bl}(r)$	bubble-diameter profile
d_S	Sauter bubble diameter
EDS	energy-dissipation spectrum
$k_L a$	volumetric mass-transfer coefficient
M_L	liquid mixing
MAB	monoclonal antibody
MW_{Pr}	molecular weight of product
MTS(r)	turbulence macro time scale profile
mm	motionless mixer
Nu	Nusselt Number (heat transfer)
OTR	oxygen-transfer rate
PI	proportional integral control
PID	proportional integral differential control
P/V	specific power input
PS	power spectrum
pH(z)	longitudinal pH profile
$pO_2(z)$	longitudinal dissolved-oxygen profile
pc	pump capacity
RTD_G	gas residence time distribution
R_X	growth rate, calculated from the OTR
SR	shear rate
SS	shear stress
T(z)	temperature profile
Tu	turbulence
Tu(r)	turbulence-intensity profile
TDT(r)	turbulence-dissipation-time profile
t_c	liquid-circulation time
X	cell-mass concentration, calculated from consumed oxygen
$w_L(r)$	liquid-velocity profile
$w_G(r)$	gas-velocity profile
$w_B(r)$	bubble-velocity distribution
ε_G	gas hold-up
η	viscosity, rheology
μ	specific growth-rate
σ_t	surface tension
σ	specific substrate-consumption rate
π	specific product-formation rate

1 Introduction

The first reports on brewing are over 5000 year old [1], but it was not until 1860 that Pasteur recognised that the alcohol was produced by living organisms in a biochemical process [2a, 2b, 2c]. In 1896, E. Buchner isolated the "fermentation" enzyme from the yeast and identified it [3]. After this time, several fermentation processes were investigated and the corresponding microorganisms were identified. Baker's yeast and fodder yeast became bulk products and were produced in submerged culture. Citric acid was originally produced in surface culture, but – later on – production was carried out in submerged culture as well [4].

However, the technology of fermentation was adapted to biochemical engineering in connection with the large-scale production of penicillin. The Waldhof-type fermenter, which was used for fodder yeast production, was successfully applied to the production of penicillin in submerged operation. Improved strains and bioreactors were developed [5–9] and advanced operation techniques were applied [10a, 10b] to penicillin production.

During the last fifty years, the biotechnology has had many highlights. Between 1950 and 1970 the main topics were the search for new antibiotics and the improvement of their production, as well as the production and biotransformation of steroids.

In order to redress the lack of proteins in developing countries, single cell protein (SCP) projects were carried out between 1970 and 1980. In western countries, yeasts were cultivated on *n*-alkanes, and – after the oil crisis – bacteria on methanol. In eastern countries, yeast was cultivated on gas oil. These projects peaked in the UK with the large-scale production of bacterial protein (Pruteen) by ICI. However, because the SCP could not compete with the inexpensive soy flour as protein fodder supplement, the projects were not economically successful.

In connection with these projects, the development of large-scale bioreactors, air-lift tower reactors in particular, were promoted.

In parallel to the SCP project, the mass cultivation of algae under non-aseptic conditions, a technology suitable for developing countries, was promoted as well. This project failed because of the resistance in developing countries to the acceptance of protein from algae.

The oil crisis between 1975 to 1985 prompted the conversion to fuel additives of renewable energy sources, such as starch, lignocellulose, and hemicellulose from plants, in addition to increased reliance on coalgas fuel. Again, large national projects for the production of ethanol and butanol were undertaken. The highlight of these projects was the production of ethanol from sugar cane in Brazil. This project too failed for economic reasons. The enzymatic decomposition of natural polymers and their conversion into solvents were also investigated in connection with these projects.

Environmental protection, especially biological wastewater treatment, was the domain of civil engineers. However, for the aerobic treatment of industrial wastewater, huge new bioreactors were developed by chemical engineers between 1975 and 1985. At the same time, biochemical engineers developed new reactors for the anaerobic treatment of heavily loaded waste-water, because the complex interaction of microorganisms in complex mixed cultures required greater knowledge of microbiology and reaction kinetics. Packed-bed- and fluidized-bed bioreactors with immobilized mixed cultures were used for this purpose.

Except for the biological wastewater treatment, the bulk-product projects were unsuccessful, because they could not compete with the low prices of the agricultural products (SCP) and of naphtha (Gasohol). Therefore, the biotechnological projects were later shifted to the development of high value products. Most of these projects were successful and initiated the development of the new industry based on the Life Sciences.

In the 1970s, projects were initiated for the production and biotransformation of secondary metabolites by plant cells (*Catharanthus roseus, Atropa belladonna, Digitalis lanata,* etc.) in cultures. However, the plant cells quickly lost their ability to form secondary metabolites in cell culture. Only few projects (e.g., shikonin) were successful. In connection with these projects the development of reactors for the cultivation of shear sensitive cells in highly viscous suspensions were promoted. The investigations with plant cells shifted later to plant breeding and the development of transgenic plants

In the1970s, insect-cell cultivation was initiated for the production of insect virus (*Autographa californica* nuclear polyhedrosis virus), which is supposed to be used as a bioinsecticide of high specificity. However, owing to its high cost, the endeavour was not realised. At present, these insect cells are becoming more widely used, mainly for the expression of high-value heterologous proteins, using recombinant baculoviruses. Insect cells are especially sensitive to shear. In connection with these projects, cell damage by shear stress and turbulence was investigated.

In 1975, Köhler and Milstein succeeded in fusing an antibody producing B-lymphocyte with a permanent myeloma cell, and were able to propagate them in a continuous culture. This success caused high activity in developing hybridoma cells and the production of various monoclonal antibodies (MABs). Because of the high demand for MABs, production was carried out in large aerated bioreactors, which had been developed especially for MAB production

Starting with naturally existing plasmids, plasmid derivatives were developed in the 1970s, and adapted to the specific requirements of genetic engineering. The construction of expression systems for the production of recombinant proteins is realized by a plasmid host system. The necessary expression-plasmids are coded for the protein product, the transcription control of which is often accomplished with inducible promoters. This development led to the start of various activities on the field of genetic engineering. The stabilization of the plasmid-carrying microorganisms had to be solved, as did the suppression of growth of the plasmid free host. The natural folding of the recombinant proteins had to be maintained. In connection with these processes, strategies were developed for the optimal induction of gene expression and for interruption of the process at the right time.

The cultivation of mammalian cells in medicine has a long story, but only the application of genetic engineering to these cells has made it possible to produce large amounts of therapeutically important post-translational modified proteins. For cultures of mammalian cells new techniques were developed: to protect the cells by low shear aeration and stirring; to reduce cost, by avoiding the use of fetal calf serum in the cultivation medium; and to increase the productivity by high cell density by means of cell-immobilization and membrane-perfusion techniques.

At the present a serious competitor is arising in the form of transgenic animals, which produce and secrete these proteins in their milk.

The formation of high value products by genetically modified microorganisms and animal cells requires highly developed process monitoring and control, in order to maintain the quality and human identity of the proteins. Monitoring the process closely allows more information to be obtained, whereupon better mathematical models are developed and better understanding of the process is gained. This is the field of modern bioreaction engineering.

Bioreaction engineering is practised mainly by chemical engineers, because chemical reaction engineering is one of its platforms [11].

The first biochemical engineering courses were organised by chemical engineering departments in MIT (Mateles et al., 1962), Columbia University, University of Illinois, University of Minnesota and University of Wisconsin in the United States, and at the University of Tokyo (Aiba, 1963) in Japan, and the first books on this subject [12–14] were published by chemical engineers and applied microbiologists [15]. After 1980, a large number of books were published on biochemical engineering (e.g., [16–26]). They provide us with a good overview of the state of the art in biochemical engineering.

2 Fundamentals

Transfer across the gas-liquid interface and mixing of the reaction components in gas-liquid chemical reactors influence the chemical reactor performance considerably. The same holds true for submerse bioreactors.

In large reactors, uniform distribution of the substrate is essential for high process performance. Aerobic microorganisms are often used for production; they have to be supplied with oxygen as well. Therefore, the fluid dynamics of the multiphase system and the transfer processes influence microbial growth and product formation. The turbulent forces, which are necessary for high transfer rate and mixing intensity, damage the microorganisms as well.

Several researchers have investigated multiphase reactors with and without microorganisms. Microbial growth and product formation were investigated in batch, fed-batch and continuous reactors, and their dependence on various parameters were described by means of mass and energy balances and kinetic equations. The reaction of the microbes to the physical and chemical variations in their environment can be explained in terms of the physiology of the microbes. Analytical methods were developed for monitoring the key parameters of the process, and the information gained is used for mathematical modelling, control, and optimization of the processes.

It is necessary to investigate the various relationships between particular variables, before the interrelationship between all of them is considered.

2.1 Fluid Dynamics and Transport Processes

In order to evaluate the interrelation between the fluid dynamics and transport processes in bioreactors on the one hand, and the microbial growth and product formation on the other, it is necessary to carry out systematic investigations with various model systems in different reactors. Fluid-dynamic investigations have mainly been performed in the chemical industry and in chemical engineering departments, with the object of designing chemical reactors, but their results are used for the design of biochemical reactors as well.

Between the first and second world wars, several large chemical companies investigated the performance of stirred tank reactors, but the results were kept secret. Only few publications dealt with this topic before and during the second world war [27–29]. In the fifties and the early sixties, several university research groups carried out similar investigations. The key issues were: power consumption, transport phenomena, mixing processes, and reactor modeling. In this period, industrial research groups were especially active, at Merck [30], du Pont de Nemours [31], and Mixing Equipment Co. [31e], all in the United States, where research in this area was also being performed at Columbia University [31c] and the Universities of Minnesota [32], Delaware [33], and Pennsylvania [8]. Similar studies were being carried out in Japan by S. Aiba at Tokyo University [34] and F. Yoshida at Kyoto University [35], in the Netherlands by van Krevelen at Staatsmijnen [36] and Kramers at TU Delft [37], and in the UK by Calderbank, in Edinburgh [38].

Later, the number of research groups dealing with multiphase reactors increased considerably (Table 1). Bubble-column- and airlift-tower loop reactors were investigated by several authors as well (Table 2). As a result, a large num-

Table 1. The leading research groups that have been dealing with fluid dynamics, transfer processes and mixing in stirred-tank reactors in the last thirty years

C.R. Wilke, H. Blanch	University of California Berkeley	USA
D.N. Miller	du Pont de Namours	USA
J.Y. Oldshue	Mixing Equipment Co	USA
F.H. Deindorfer	University of Pennsylvania	USA
M. Moo-Young,	University of Waterloo	Canada
C.W. Robinson	University of Waterloo	Canada
J. Carreau	Ecol. Poy.Techn. Montreal	Canada
A.W. Nienow	University of Birmingham	UK
J.J. Ulbrecht	University of Salford	UK
H. Angelino, J.P Courdec	CNRS, Toulouse	France
H. Roques, M. Roustan	INSA, Toulouse	France
J.C. Carpentier	CNRS Nancy	France
A. Mersmann	University of Munich	Germany
U. Werner, H. Höcker	University of Dortmund	Germany
P.M. Weinspach	University of Dortmund	Germany
H. Brauer	TU Berlin	Germany
M. Zlokarnik, H.J. Henzler	Bayer Co.	Germany
H. Kürten, P. Zehner	BASF Co.	Germany
H.Ullrich	Hoechst Co.	Germany

Table 1 (continued)

K.D. Kiepke	EKATO Rühr u. Mischtechnik	Germany
F. Liepe	Inst. F. Strömungstechnik, TU Köthen	E. Germany
F. Yoshida	University of Kyoto	Japan
J. Kobayashi	University of Tsukuba	Japan
T. Kono	Takeda Chem. Ind. Co	Japan
J.M. Smith	TU Delft	Netherlands
D. Thoenes	University of Twente	Netherlands
K. van't Riet	University of Wageningen	Netherlands
J. van de Vusse	Koninklijke Shell Co	Netherlands
A. Fiechter	ETH Zurich	Switzerland
V. Linek	Chem. Techn. Inst. Prague	Czechoslovakia
U.E. Viesturs	Latvian Acad. Sci. Riga	Latvia
M. Raja Rao	IIT Bombay	India

Table 2. The leading research groups that have been dealing with bubble column- and airlift-tower-loop reactors in the last thirty years

Y.T. Shah	Pittsburgh University	USA
M.L. Jackson,	University of Idaho	USA
D.N. Miller	du Pont Namours	USA
G.A. Hughmark	Ethyl Co. Baton Rouge	USA
J.R. Fair	Monsanto Co.	USA
M. Moo-Young, Y. Chisti	University of Waterloo	Canada
C.W. Robinson	University of Waterloo	Canada
M.A. Bergougnou	University of Western Ontario	Canada
H. Kölbel	TU Berlin	Germany
H. Hammer	TH Aachen	Germany
H. Langemann, H.J. Warnecke	University of Paderborn	Germany
W.D. Deckwer, A. Schumpe	University of Hannover	Germany
	University of Oldenburg, GBF	Germany
H. Blenke	University of Stuttgart	Germany
U. Onken, P. Weiland, R. Buchholz	University of Dortmund	Germany
K. Schügerl	University of Hannover	Germany
A. Vogelpohl, N. Räbiger	TU Clausthal	Germany
W. Sittig, W.A. Stein, L. Friedel	Hoechst Co	Germany
H. Zehner	BASF Co	Germany
M. Zlokarnik	Bayer Co	Germany
J.F. Davidson	University of Cambridge	UK
J.F. Richardson	Imperial College London	UK
E.L. Smith, N. Greenshields	University Aston, Birmingham	UK
J.S. Gow, J.D. Littlehails	ICI, Billingham	UK
J. Tramper, K. van't Riet	University of Wageningen	Netherlands
J.J. Heijnen	Gist brocades/TU Delft	Netherlands
Y.F. Yoshida	University of Kyoto	Japan
T. Miauchi	University of Tokyo	Japan
Y. Kawase,	Toyo University	Japan
T. Otake,	Osaka University	Japan
Y. Kato, S. Morooka	Kyushu University	Japan
J.B. Joshi, M.M. Sharma	IIT Bombay	India
J.C. Merchuk	Ben Gurion University	Israel
F. Kastanek	Inst. Proc. Fund., Prague	Czechoslovakia

ber of experimental data in laboratory scale are at our disposal, which allow, for example, the prediction of mixing times and oxygen-transfer rates. However, data for large-scale reactors are still scarce. The results of these investigations are summarized in several books [21, 39, 40]. Stirred-tank reactors have recently been modeled with Computational Fluid Dynamics (CFD) [41–45]. Bubble column reactors were modeled with CFD by solving the Navier-Stokes Differential-Equation System [46–51]. These calculations offer greater insight into the fluid dynamics and transfer processes.

2.2 Macroscopic Total Mass, Elemental Mass, Energy and Entropy Balances

Interrelations between the rates of growth, product synthesis, respiration, and substrate consumption have been studied by the macroscopic balance method. Minkevich and Eroshin [52] developed the degree of reduction concept, which considers the number of electrons available for transfer to oxygen combustion.

Erickson [53], Roels [54], Stouthamer [55], and Yamané [56] have further improved this concept. This method was applied on several biological systems (Table 3). The macroscopic balances provide useful relationships for the analysis of growth and product formation. They allow the prediction of the yield coefficients and efficiency factors, e.g. with different electron acceptors.

2.3 Kinetics of Growth and Product Formation

The early investigations of bacterial growth kinetics were reviewed by Hinshelwood [81]. Empirical investigations indicated that the dependence of cell growth on substrate concentration is the same as that of enzyme kinetics, in which Michaelis-Menten kinetics [82] is generally accepted, and which had been extended to competitive and non-competitive inhibitions and complex enzymatic reactions [83].

Table 3. Application of macroscopic balances to various biological systems

Bakers yeast	[54c, 57–63]
Penicillium chrysogenum	[54c, 64, 65]
Candida utilis	[66]
Escherichia coli	[67–69]
Rhodopseudomonas sphaeroides	[70]
Tetracycline by *Streptomyces aureofaciens*	[71]
Gluconic acid by *Aspergillus niger*	[72]
Poly-β-hydroxy-butyric acid by *Alcaligenes eutrophus*	[73]
Klebsiella pneumoniae	[74]
Conversion of D-xylose to 2,3 butanediol by *Klebsiella oxytoca*	[75]
Enterobacter aerogenes	[76]
Paracoccus denitrificans	[77]
Propionibacterium	[78]
Several microorganisms	[54c, 79, 80]

Monod recommended an analogous relationship for bacterial growth [84], and applied it to several biological systems. The Monod equation was then extended to special cases of bacterial growth, and relationships were developed to cover product formation as well [85–91]. Continuous cultivation of microorganisms became popular. Mass-balance relationships for steady state and substrate limited cultivation (Chemostat) were published [92–102]. These relationships were used for macroscopic material balances in cultures [54c–80].

2.4 Metabolic pathways

A large number of researchers have participated in the discovery of the metabolic pathways of living cells. In the 1930s and 1940s, the glycolysis and the tricarboxylic acid cycle were recognised [103–106]. Overviews of these investigations were presented in the 1950s and 1960s [107, 108]. The present state of the art has been described by Doelle [109] and by Gottschalk [110]. The results of these investigations, and of careful measurements of the concentrations of the main components during the cultivations, allow quantitative analysis of the metabolic fluxes.

2.5 Process Monitoring and Control

2.5.1 *pO_2 and pH Measurement*

Since its introduction by Clark [111], the membrane-covered dissolved oxygen electrode and its modified versions have been used widely in the practice of biotechnology. The pH-electrodes with glass membrane are based on investigations of MacInnes and Dole [112]. These sodium-glass membranes are still manufactured and sold under the designation CORNING 015, but modern pH glasses contain lithium oxide instead of sodium oxide and have a much wider measuring range [113].

Temperature, dissolved oxygen, and pH are measured in-situ; the other key process variables are monitored either off-line or on-line.

2.5.2 *Biosensors*

Biosensors are especially suitable for the analysis of complex culture media. They consist of a chemically specific receptor and a transducer, which converts the change of the receptor to a measurable signal. Enzymes, cells, antibodies, etc., are used as receptors. A good review of the history of biosensor development is given in the book of Scheller and Schubert [114]. Enzymes have been used as early as 1956 for diagnostic purposes. The first transducer was a pH sensor combined with phosphatase [115]. The oxygen sensor was first used by Clark and Lyons [116] as the transducer in combination with glucose oxydase

(GOD). Updike and Hicks were the first to immobilize a (GOD)-receptor in a gel. [117]. Enzyme electrodes were also developed by Reitnauer [118]. The first analytical instrument with immobilized enzyme was Model 23 A was put on the market by Yellow Springs Laboratory [119]. Lactate analyzer 640 La Roche was the next commercial instrument [120]. The first enzyme-thermistor was developed by Mosbach [121], and Loewe and Goldfinch [122] developed the first optical sensor. A bacterium was used as receptor instead of enzyme for alcohol analysis by Divies [123]. Cell organelles were used by Guibault for NADH analysis [124], and synzymes by Ho and Rechnitz [125]. Antibodies were introduced by Janata [126] and receptor proteins by Belli and Rechnitz [127] for biosensors. In the last 15 years, the different types of biosensors were being developed [128]. Their application is restricted to laboratory investigations. They are often used in flow injection analysis (FIA) systems as chemically specific detectors [129, 130]. A short analysis time is a prerequisite for process control. Flow-injection analysis, with response times of few minutes, is especially suitable for on-line process monitoring. Flow-injection analysis, developed by Ruzicka and Hansen [131], became popular in the last twenty years in both chemistry [132] and biotechnology [133].

2.5.3
On-line Sampling, Preconditioning and Analysis

The prerequisites of on-line process monitoring are aseptic on-line sampling, sample conditioning, and analysis. The first on-line sampling systems used a steam flushed valve system, consisting of a sampling transfer-tube from the reactor to the analyser, steam supply, a condenser, and four valves for successively sterilizing the transfer tube, withdrawing the sample, and cleaning the transfer tube. Such systems were used for on-line sampling by Leisola et al. [134, 135]. The medium losses, which were considerable, were reduced by miniaturization [136, 137]. Dialysers were the first cell-free sampling systems [138, 139, 140]. Later on, UF membrane filtration was used for sampling and analysis of low molecular-weight analytes, and MF membrane filtration for sampling and analysis of proteins. The first external cross-flow aseptic membrane module that was integrated into a medium recirculation loop [141] was commercialized by B. Braun Melsungen (BIOPEM®); another system [142] was produced by Millipore. The first internal in situ filter for sampling [143] was commercialized by ABC Biotechnologie/Bioverfahrenstechnik GmbH. A coaxial catheter for cell-content sampling was developed by Holst et al. [144], but it was not commercialized. For gas sampling, silicon-membrane modules can be used [145].

Sample conditioning for the analysis of low-molecular-weight components of the medium consists of cell removal, protein removal, dilution or enrichment of the analytes, correction of pH and buffer capacity, removal of toxic components and bubbles, degassing the sample, suppression of cell growth by growth inhibitors, etc. [146].

Modern on-line monitoring systems offer automated sampling, sample conditioning, and analysis [147–149]. Short sampling-, preconditioning-, and analysis times are prerequisites for process control. The internal in situ sampling system and flow injection analysis with response times of few minutes are especially suitable for on-line process monitoring. On-line gas chromatography (GC) [150, 151, 152] and high performance liquid chromatography (HPLC) [153, 154] are used for process monitoring as well, but their analysis times are several minutes. Mass spectrometry is used for in-line off-gas analysis [155]. Lately, in situ process monitoring with near-infrared Fourier transform (NIR-FT) spectroscopy [156] and 2D-fluorescence spectroscopy [157] became possible. The present state of bioprocess monitoring has been described by Schügerl [130, 158].

2.5.4 *Process Control*

The classical low-level automatic controls include analog-, on off, sequence-, and feedback control [159, 160].

Low-level controllers are used for the control of flow (PI), gas pressure (PI), temperature (PID), rotational speed of the stirrer (PI), pH-value, dissolved oxygen (PI) and sequence (sterilization, batch and fed-batch process).

Modern control theory was developed between 1950 and 1960 and was applied in biotechnology in the 1970 s. At the same time, advanced computer hardware, especially microcomputers were being developed. Pioneers in computer control were Armiger and Humphrey [161], Bull [162], Hampel [163], Hatch [164], Jefferis [165], Lim [166], Weigand [167], and Zabriskie [168]. The development of computer control is well represented by presentations in the Congresses on Computer Application in Biotechnology [169–173]. The state of the art of control of bioreactor systems has been described by Wang and Stephanopoulos [174], by Lim and Lee [175], and by Bastin and Dochain [176].

2.6 Mathematical Models

The earliest models related growth to the growth-limiting substrate [85, 177–182], and were extended by including inhibition kinetics. (For a review, see Reference [183].)

Later, models of cell population with segregated and structured models were been developed. Tsuchiya et al. [184] classified the mathematical models of microbial populations according to Fig. 1. Segregated models consider the heterogeneity of individuals, whereas structured models take the various cell components into account.

Ramkrishna et al. [185, 186] introduced *cybernetic modeling*, which assumes that the cells choose the possible pathways that optimize their proliferation. Shuler et al. [187, 188] developed large-scale computer models for the growth of a single cell. Other structured cell models, developed by Perretti and Bailey [189, 190], take into account the perturbation of the metabolism that occurs as a result of the introduction of recombinant plasmids. The genetically structured models of Lee and Bailey [191] consider plasmid replication in recombinant microorganisms. Other models deal with the proliferation rate influenced by

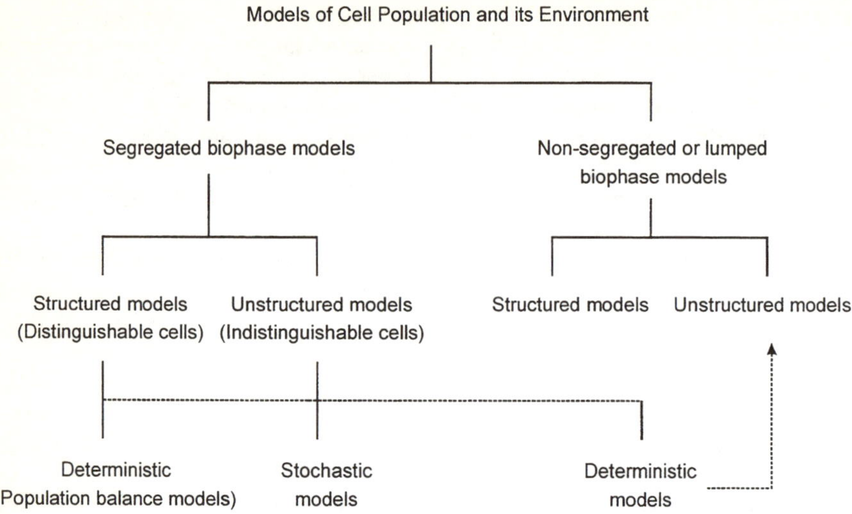

Fig. 1. Classification of mathematical models of microbial population [144]

exogenous growth factors [191]. The metabolic engineering models, recommended by Bailey [192], use the known stoichiometric structure of the intracellular reaction network by assuming a quasi-steady state of the intermediate intracellular metabolites, in order to obtain intracellular fluxes.

However, for process optimization and control, simple structured (so-called compartment) models are used. Harder and Roels compiled common two- and three-compartment models in their review [193].

Several mathematical models, based on the cellular regulation model of Jacob and Monod [194], were developed for the genetic control of enzyme synthesis. These publications were reviewed by Harder and Roels [193]. A typical process model was presented by Bellgardt [195].

3 Interrelation Between Physical, Chemical and Biological Processes

The prerequisites for the determination of the interrelation between physical chemical and biological processes are:

1) closely monitored and controlled cultivation;
2) monitoring of the key fluid-dynamic properties;
3) monitoring of the concentrations of the key medium components;
4) monitoring of the concentration and biological state of the cells.

Very few investigations are known that fulfil all of these essentials, but several have been published that satisfy two or three of them.

3.1 Influence of Fluid Dynamics and Transport Processes on Microbial Cultures

Most of the microbial cultivations are performed with monocultures. The important prerequisite is a monoseptic operation, using sterile medium and avoiding infection during the cultivation. The sterility of the bioreactors, necessary for the large-scale production of penicillin, was accomplished by the development of suitable rotating seals for the stirrer shaft. The connections, which are necessary for the fluid-dynamic and mass-transfer measurements, impair the performance of the process and the sterility of the system. Therefore, special setups and runs are necessary for the evaluation of these properties.

The determination of the specific power input (P/V) is only possible for large reactors, because instruments for torque measurement with torsion dynamometer and strain gauges are on the market only for large stirrers. Power monitoring with a wattmeter is only accurate for large reactors, for which the power uptake by frictional losses at the rotating seals of the stirrer shaft are negligible in comparison with the power uptake by mixing and gas dispersion. Therefore, for investigations in laboratory stirred-tank (ST) reactors, power-input data are not available. The specific power input can be calculated from the aeration rate in bubble-column (BC)- and airlift-tower-loop (ATL) reactors. The measurement of the gas hold-up in ST reactors is difficult. In BC and ATL reactors, it can be calculated from the pressure difference between the bottom and the headspace, and by monitoring the liquid level with capacity sensors. Steel and Maxon [196–198] performed the first systematic investigation of the influence of specific power-input on fermentation performance. They investigated the production of Novobiocin by *Streptomyces niveus* in stirred-tank reactors of different capacity (20 l, 250 l, 3000 l, and 6000 l) and with various impellers [196–198]. Their main interest was the dependence of gas hold-up (ε_G), oxygen transfer rate (OTR), and productivity (Pr) on the specific power input (P/V), speed (N), and diameter (d_N) of the impeller, the aeration rate (Q_G) and the viscosity of the culture medium (η). In a review, Cooney and Wang compared the OTR and OTR-efficiencies – with regard to power input – of yeast, *Endomyces*, and *Streptomyces*) in cultures in different industrial reactors varying in volume form 30 to 128 m^3 [199].

Based on the early investigations, the Rushton impeller became the standard stirrer in biotechnology. Only recently, new impellers such as the Scaba agitator and hydrofoil agitators (Lightnin A315 and Prochem Maxflow T), with higher mixing and oxygen transfer efficiencies, have come into used (see [200, 201]).

In the early days of bioprocess technology, bubble column reactors were preferred, because it was easier to maintain their sterility. After the sterile rotation seal for stirrer shaft was developed, ST reactors became the standard reactors for industrial production, because of their flexibility and high performance, especially for highly viscous culture media. AS reactor size increased, ST reactors were replaced by bubble-column (BC)- and airlift-tower-loop (ATL) reactors, mainly in the aerobic wastewater treatment plants of chemical factories. Comparison of these reactor types indicated that ST is a high-performance reactor with low OTR-efficiency with regard to the power input, whereas BC and ATL are medium performance-reactors with high OTR-efficiency with regard to the power input [202–205]. Therefore, less heat is evolved during the operation in BC and ATL reactors than in ST reactors.

Fiechter and Adler [206, 207] compared the performances of compact stirred-loop (CSL)-reactors with overall volume of 50 l, 550 l, an ATL reactor with 2300 l ATL overall volume, a 100 l torus reactor, and 7 l, 30 l standard STs, by cultivating the yeast *Trichosporon cutaneum*,, which is insensitive to glucose repression and does not produce ethanol under oxygen limitation. Therefore, there is a direct relationship between the rates of growth and of oxygen uptake. The growth rate can be calculated from the OTR, and the cell concentration from the consumed oxygen. Xanthan production by *Xantomonas campestris in a* highly viscous medium [204, 208, 209] and *T. cutaneum* cultivation in a medium of low viscosity [210, 211] were used for the comparison of the performances of 30 l, 1200 l ATL/BC bioreactors with 15 l, 300 l, 1500 l, 3000 l ST reactors by Deckwer et al [208–211]. They recommended relationships for the calculation of the volumetric mass transfer coefficient (k_La) as well. Several investigations were carried out with other microorganisms (Table 4). In Table 4

Table 4. Fluid-dynamic investigations with microbial cultivation systems. Stirred tank (*ST*), Airlift-tower-loop (*ATL*)-, Bubble-column (*BC*)-, compact-stirred-loop (*CSL*)-, and stirred-loop (*SL*) reactors

Organisms	Reactor	Volume (L)	Measurements	Ref.
Trichosporon cutaneum	ST	7, 30	P/V, OTR, R_X	[206, 207]
T. cutaneum	CSL	50, 550	P/V, OTR, R_X	[206, 207]
T. cutaneum	ATL	2300	P/V, OTR, R_X	[206, 207]
T. cutaneum	Torus	100	P/V, OTR, R_X	[206, 207]
T. cutaneum	BC/ATL	30, 1200	OTR, k_La, R_X	[210]
T. cutaneum	ST	300, 1500, 3000	OTR, k_La, P/V	[211]
Xanthomonas campestris	BC	13	k_La	[203]
X. campestris	ATL/BC	50, 1200	k_La, OTR, MW_{Pr}, η	[204]
X. campestris	ST	1500	k_La, OTR, MW_{Pr}, η	[204]
X. campestris	ST	15, 100	OTR, MW_{Pr}, η	[208]
X. campestris	ST	15, 100	k_La, η, P/V	[209]
X. campestris	SL	900	k_La, η, P/V	[209]
X. campestris	ST	10	OTR, SR, SS	[212]
Escherichia coli	ATL	60	M_L, pO_2, OTR, k_La,	[205]
E. coli	ST	10	OTR, k_La	[205]
E. coli	ATL	100	$pO_2(z)$, $d_{Bl}(r)$, d, $\varepsilon_G(r)$, OTR, $k_La(r)$, a, $w_L(r)$, $w_{Bl}(r)$, Tu(r), MTS(r), PS, EDS, TDT(r)	[213]
Saccharomyces cerevisiae	ATL	4000	$w_L(r)$, RTD, t_c, D_L	[214]
S. cerevisiae	ATL	250	ε_G, t_c	[215]
S. cerevisiae	ATL	4000	RTD	[216]
S. cerevisiae	ATL	4000	RTD_G, w_L, D_L, D_G	[217]

Table 4 (continued)

Organisms	Reactor	Volume (L)	Measurements	Ref.
S. cerevisiae	ATL	80, 4000	RTD_G, w_L, ε_G, D_L, D_G, w_B, (in riser and down-comer)	[218a, b, c]
Gluconobacter oxydans	ST	3.2, 90	M_L	[219]
G. oxydans	ST	25,000	t_c, pc, k_La, pO_2	[220, 221]
Methyphilus methylotrophus	ATL	30	M_L, t_c	[222]
Candida utilis	BC	20	a	[223]
Leuconostoc mesenterius	BC	20	a	[223]
Penicillium chrysogenum	ST	7, 15, 41	M_L, η	[224]
P. chrysogenum	ST	1500	RTD_G	[225]
Aspergillus niger	ST	50	OTR	[226]
Bacillus subtilis	ST	14	pO_2 (< 100 ppb) butanediol, pO_2 (> 100 ppb) acetoin	[227]
Neurospora sitophila	ATL	1000	t_c, ε_G, h	[221]
Chaetomium cellulolyticum	BC	40	t_c, ε_G, h	[228]
C. cellulolyticum	ATL	1300	ε_G, η, OTR, k_La	[229]
N. sitophila	ATL	1300	ε_G, η, OTR, k_La	[229]
C. cellullyticum	BC	40	k_La, η, σ, a (calc.)	[230]
Kluyveromyces fragilis	ATL	120,000	k_La, OTR, pO_2,	[231]
K. fragilis	ST	15, 100	OTR	[232a, b, c]
Hansenula polymorpha	ATL	60	$pO_2(z)$, , OTR, k_La, a, d_{Bl}, d_S	[232a, b, c]
H. polymorpha	ATL	60	$pO_2(z)$, $d_{Bl}(r)$, d_S, $\varepsilon_G(r)$, OTR, k_La, a, $w_L(r)$, $w_{Bl}(r)$, Tu (r), MTS (r), PS, EDS, TDT (r)	[213]
Streptomyces niveus	ST	20, 250, 15,000	P/V, OTR, Pr	[196, 197, 198]
S. aureofaciens	ST	112,000	pO_2, η, T, k_La	[233]
S. avermitilis	ST	800	k_La, OTR, P/V	[200]
S. rimosus	ST	20, 60	Nu, η	[234]
S. albus	ST	20, 60	Nu, η	[234]
Streptomyces	BC/ATL	20,000	M_L, t_c, k_La, h, P/V, pO_2, pH	[235]
Bacillus licheniformis	ST	100, 67,500	k_La, η, P/V	[236]
Endomyces sp.	ST	20, 40	P/V, η	[237]
Aureobasidium pullulans	ST	8, 14	DOC	[238]
Candida boidinii	ATL	60	ε_G, d_{Bl}, d_S, OTR, k_La, a, P/V, Pr, pO_2, (concurrent and counter-current operation)	[239]
cell free system	ATL (m.m)	60	w_L, D_L, w_B (small, large horizontal, vertical)	[240]
cell free system	ATL	160	$w_L(r)$, $w_G(r)$, $\varepsilon_G(r)$,	[241]

only investigations are listed that were performed with stirred-tank-, bubble-column-, and airlift-tower-loop reactors with volumes of at least 20 l, because the fluid dynamics in smaller reactors considerably differs from those in large ones. In most of these investigations, the temperature, (T), the stirrer speed (N), the aeration rate (Q_G), and the pH-value were controlled, and the dissolved-oxygen concentration (pO_2) was monitored. In addition to the concentrations of the cell mass, the substrate (S), and the product (P), the composition of off-gas (O_2, CO_2) were measured. Based on the mass balances, these data allow the evaluation of the specific growth rate (μ), the specific substrate-uptake rate (σ), and specific product-formation rate (π), as well as the volumetric mass-transfer coefficient (k_La), the rates of carbon dioxide formation (CPR) and oxygen uptake (OUR) and their ratio – the respiratory quotient (RQ). The yield coefficients of growth and product formation with regard to the substrate consumption ($Y_{X/S}$ and $Y_{P/S}$) are calculated as well. These process variables were determined in several modern fluid-dynamic investigations that were performed during microbial cultivation. The relationships, which were developed on the basis of these investigations, differentiate between low viscous cultivation media (e.g., *T. cutaneum*) [211] and highly viscous non-Newtonian cultivation media (e.g., *Xanthomonas campestris*) [209]. However, this classification only holds true for systems in which the high viscosity is caused by the product (e.g., xanthan), and not by the microorganisms (fungi or streptomycetes). In the latter case, the cell morphology has a considerable influence on the interrelationship between fluid dynamics and transport processes, and – thus – on cultivation performance. Therefore, no generally applicable relationships exist.

With increasing reactor size, cell concentration, and medium viscosity, the distributions of substrate and dissolved oxygen in a reactor becomes less and less uniform. In large ST reactors, three relatively distinct regions with different mixing intensities exist: (a) a well mixed impeller region (micro-mixer); (b) a slightly mixed bulk region (macro-mixer); and (c) a wall region (dead water). The microorganisms circulate through the micro-mixer – with high substrate and dissolved oxygen concentrations, and the macro-mixer – with low substrate and dissolved oxygen concentrations. Depending on the circulation-time distribution, the microorganisms can become substrate and oxygen limited, respectively, in the macro-mixer region. Bajpai and Reuss considered the coupling of mixing and microbial kinetics in different reactors [226]. In ATL reactors, only the riser is aerated, therefore, oxygen limitation can occur in the down-comer, again depending on the circulation-time distribution in the reactor.

To investigate the dynamic behaviour of the cells in periodically varying environmental conditions, several authors cultivated yeast cells in a small reactor varying the dissolved oxygen concentration periodically and circulating the cell suspension through aerated and nitrogen gassed small reactors, respectively, or adding glucose pulse to the reactor and monitoring the concentrations of cell-mass, ethanol, dissolved oxygen, and NAD(P)H-dependent culture fluorescence [242–249]. The measurements in industrial Baker's yeast ATL reactors indicated that the ethanol, which was produced by the yeast in the down-comer, was consumed in the riser, as long as the volume ratio of riser to down-comer was large enough [250, 251]. However, in large ST reactors with highly viscous media this phenomenon impairs reactor performance considerably [226].

3.2
Process Identification by Advanced Monitoring and Control

The regulation of metabolic flux and intracellular metabolite concentrations is exerted at various levels, which can be roughly divided into two categories: (a) modulation of enzyme specific activity or affinity to substrates; and (b) modulation of enzyme concentration. Typical examples of the first category include cooperative effects, allosteric effects, and covalent modifications, whereas the second includes transcriptional and translational control [252]. Metabolic control analysis (MCA) had already been developed in the 1970 s [253–256], but the practical application of these results was hindered by experimental difficulties. Twenty years later, the experimental techniques attained the necessary level for the application of MCA. In addition to the standard process variables (T, N, Q_G, pH, X, Pr, S, pO_2, O_2, CO_2), several other variables have to be monitored for metabolite control of growth and product formation. Such variables are the concentrations of phosphate, ammonium, amino acids, primary metabolites (acetate, lactate, pyruvate, succinate, ethanol, etc.), the precursors of the secondary metabolites, NAD(P)H, RNA, DNA, proteins, and cell morphology [130, 257]. The heat generated by the microorganisms [258–263] and the intracellular enzyme activities and metabolites [264–269] are monitored as well. In special cases, the intracellular enzyme activities were measured on-line [270–274].

Using advanced measurement and control techniques in well-mixed reactors, highly reproducible data were obtained[275], allowing pathway analysis in microorganisms, e.g., glucose transport [276], oxygen utilization, and the determination of tricarboxylic acid cycle activity [277]. On-line process identification, by means of elemental and macroscopic balancing and advanced data processing, thus became possible [68, 69, 166, 278, 279].

3.3
Metabolic Engineering, Metabolic Flux Analysis

The regulation of metabolic networks is complex, because they involve several enzymes and a great variety of control mechanisms. Quantitative analysis of intracellular fluxes is an important tool for investigating control mechanisms.

Intracellular fluxes can be determined by measuring the rates of change of extracellular metabolite concentrations and using the total mass and carbon balance to calculate them. The determination of intracellular enzyme activities and metabolite concentrations can supplement these measurements. Particular fluxes along some pathways can be measured directly by NMR spectroscopy using ^{1}H and ^{13}C-, or ^{31}P-isotopes. However, direct flux measurements with NMR is impaired by insensitivity. The combination of the mass balances and NMR spectroscopy allows the identification of the critical junctions (nodes) in a network that influence the partitioning of the fluxes, and the intracellular measurements permit the determination of the type of enzymatic modification. In Table 5, several investigations are compiled.

The extension of metabolite balancing with carbon-isotope labeling experiments allows the quantitative determination of the flux of bidirectional reactions in both directions [306]. The simultaneous application of flux balancing, fractional ^{13}C-labeling of proteinogenic amino acids and two-dimensional NMR-spectroscopy, as well as automatic analysis of the spectra, provides a rapid, double checked analysis of the fluxes [306]. The application of the combination of these techniques led to important new results on the

Table 5. Metabolic flux, metabolic pathway investigations (The NMR-measurements were combined with mass balancing)

Organism	Method	Product	Reference
Bacillus subtilis	mass balancing		[280]
B. subtilis	2D [^{1}H, ^{13}C]-NMR	riboflavin	[281]
Corynebacterium glutamicum	in-vitro-^{13}C NMR	lysine	[282]
C. glutamicum	mass balancing	lysine	[283]
C. glutamicum	in-vitro-^{13}C NMR	lysine	[284]
C. glutamicum	mass balancing	lysine	[285]
C. glutamicum	in-vivo-^{1}H, ^{13}C NMR	lysine	[286]
C. melassecola	mass balancing	glutamic acid	[287]
Escherichia coli	mass balancing	(acetate)	[288]
E. coli	mass balancing	(acetate)	[289]
Hybridoma	in-vivo-^{13}C NMR	MAB	[290]
Hybridoma	in-vitro-^{13}C NMR	MAB	[291]
Hybridoma	mass balancing	MAB	[256, 292]
Penicillium chrysogenum	mass balancing	penicillin G	[293]
P. chrysogenum	mass balancing	penicillin G	[258, 294]
P. chrysogenum	mass balancing	penicillin G	[295]
Saccharomyces cerevisiae[a]	in-vivo-^{31}P NMR	(ethanol)	[295]
S. cerevisiae	mass balancing intracell. comp (ethanol)		[297]
S. cerevisiae	mass balancing intra cell. comp. (ethanol)		[298]
S. cerevisiae	mass balancing intra cell comp. (ethanol)		[299]
S. cerevisiae	mass balancing	(ethanol)	[300]
S. cerevisae	mass balancing	(ethanol)	[301]
Spodoptera frugiperda	mass balancing	DNA, RNA, protein	[302]
Streptococcus lactis	mass balancing	lactic acid	[303]
Zymomonas mobilis	mass balancing intracell. enzyme activity ethanol		[304]
Z. mobilis	in-vivo-^{31}P NMR	ethanol	[305]
Z. mobilis	in-vivo-^{1}H, ^{13}C NMR	ethanol	[306]

[a] immobilized cells

distribution of metabolic flux in the cells. Recent results on metabolic engineering are reported in a special issue of Biotechnology Bioengineering [307].

3.4 Expert Systems, Pattern Recognition

Expert systems form a special class of artificial intelligence (AI) [308–310]. They handle problems by reasoning, drawing on a knowledge-base generated by human experts who are well versed in the problem in question [311]. Expert systems have been used in chemical engineering [312, 313] and biochemical engineering [314–339]. In Table 6, several biotechnological applications of expert systems are compiled that combine mathematical models with additional heuristic knowledge. Biotechnological processes are highly non-linear and – especially – non-stationary. Hence, their dynamic characteristics change with the time and, consequently, the parameters of the model vary during cultivation. A common way of coping with this problem is to use less detailed models and parameter-adaptive techniques to periodically fit these models to the actual process data. Another way of solving this problem is to use a distributive model – consisting of several sub-models – by which the accuracy of the Extended Kalman Filter used for the estimation of the procedure can be improved, and fuzzy reasoning techniques to incorporate additional heuristic knowledge about the process [332]. By using a modular set of artificial neural networks (ANN), it is possible to obtain reliable real-time state estimations and process predictions [333].

Pattern recognition is used to reconstruct characteristics of the system that are only partially known. Patterns in technological processes are defined as a

Table 6. Application of expert systems

Biological system	References
Baker's yeast cultivation	[314–316]
Bacillus subtilis (α-amylase production)	[317, 318]
Escherichia coli cultivation	[319–323]
Escherichia coli (α-interferon production)	[324]
Lactobacillus casei (lactic acid production)	[325, 326]
Zymomonas mobilis cultivation	[327]
Penicillium chrysogenum (penicillin G production)	[328, 329]
Antibioticum production	[330]
Bacillus amyloliquefaciens (α-amylase production)	[331]
Glucoamylase production	[327]
Fusarium graminearum cultivation	[328]
Beer production	[332–334]
Enzyme production	[335]
Mammalian cell cultivation	[336, 337]
Hybridoma cell (monoclonal antibody production)	[338]
CHO-bioreactor development	[324]
α-interferon recovery and purification	[324]
Pilot plant operation in Eli Lilly Co.	[339]

combination of signals transmitted by different sensors. Each single trajectory occupies a row in the matrix. Hence, this results in as many rows as there are sensors, and as many columns in each row as there are discretized time-steps [342–346]. Pattern recognition provides the possibility of distinguishing many different actual states of a culture, making it a suitable tool for expert systems.

4 Particular Systems

The bioreaction-engineering problems that arise in connection with immobilized cells and high-density cell cultures, as well as with animal and plant cells, differ considerably from those of low-density microbe cultures. Therefore, these systems are considered separately in this section.

4.1 Immobilized Micro-organisms

Microorganisms can be immobilized by growing them on solid support (biofilm formation), by adsorption or colonization in porous supports, or by entrapment within gels, such as synthetic polymers, kappa carrageenan, agarose, alginate, chitosan, cellulose acetate, and collagen. Cell immobilization allows high cell density to be obtained, and high dilution rates can be applied in continuous cultures without cell washout. However, internal gradients in the microbial aggregates can reduce process performance and gas evolution can destroy the gel beads. Biofilms have been used for acetic acid production for a long time. Enzymes were first immobilized by Manecke in 1959 [347] and by Katchalski in 1964 [348]. Mosbach reported the entrapment of enzymes and microorganisms in cross-linked polymers in 1966 [349], and Johnson entrapped fungal spores in solid matrices in 1969 [350]. Mosbach et al. used immobilized micro-organisms for steroid conversion in 1970 [351]; Toda and Shoda used immobilized yeast cells for sucrose inversion in 1975 [352]; Kiersten and Bucke immobilized microorganisms in 1977 [353]; and Fukui and coworkers immobilized microbodies for biosynthesis in 1978 [354]. A review of the early work on immobilized enzymes and microorganisms was published by Wada et al. in 1979 [355] and by Fukui and Tanaka in 1982 [356]. Reviews on immobilized enzymes were presented in 1978/79 [357–359].

Subsequently, several publications appeared on immobilization of enzymes and microorganisms. The early publications on immobilized microorganisms are listed in Table 7. The number of research groups has working in this area has increased since the 1980s. Furusaki and Seki reviewed the reports published in 1990/91 on immobilized microorganisms and on animal and plant cells.

The first industrial application of immobilized enzymes was presented by Chibata in 1978 [357]. Shibatani reported on the application of immobilized biocatalysts in Japan in 1996 [395].

A central interest in reaction engineering is intraparticle diffusion of substrates, which has been identified as the rate-limiting process. The earliest investigations of substrate-diffusion in spherical gel-particles, using immobilized

Table 7. Early publications on immobilized microorganisms

Microorganisms	Product	Year	Reference
Acetobacter xylinum	glycerol → dihydroxy acetone	1979	[360]
Acetobacter suboxydans	acetic acid	1977	[361]
Acetobacter cells	acetic acid	1982	[362]
Brevibacterium flavum	L-malic acid	1980/83	[363, 364]
Gluconobacter oxydans	gluconic acid	1981	[365]
Brevibacterium flavum	glutamic acid	1981	[366]
Brevibacterium ammoniagenes	L-malic acid	1976	[367]
Pseudomonas dacunhae	L-alanine	1980	[368]
Brevibacterium ammoniagenes	NADP	1979	[369]
Serracia marcescenses	isoleucine	1980	[370]
Achromobacter liquidum	urocanic acid	1974	[371]
Saccharomyces cerevisiae	ethanol	1981	[372]
Saccharomyces cerevisiae	ethanol	1982	[373]
Saccharomyces cerevisiae	ethanol	1980	[374]
Saccharomyces cerevisiae	ethanol	1983	[375]
Saccharomyces cerevisiae	ethanol	1981	[376]
Saccharomyces cerevisiae	ethanol	1981	[377]
Saccharomyces cerevisiae	ethanol	1981	[378]
Zymomonas mobilis	ethanol	1983	[379]
Zymomonas mobilis	ethanol	1981	[380]
Pachysolen tannophilus	D-xylose → ethanol	1982	[381]
Streptomyces catteya	thienamycine	1983	[382]
Escherichia coli	D-tryptophan	1979	[383]
Penicillium chrysogenum	Penicillin G	1984	[384]
Penicillium chrysogenum	Penicillin G	1979	[385]
Bacillus sp.	Bacitracin	1980	[386]
Candida tropicalis	phenol decomp	1979	[387]
Pseudomonas denitrificans	dentritification	1980	[388]
Bacillus subtilis	Proinsulin	1983	[389]
Bacillus subtilis	α-amylase	1978	[390]
Streptomyces fradiae	protease	1981	[391]
Claviceps purpurea	clavine alkaloids	1983	[392]
Claviceps purpurea	clavine alkaloids	1982	[393]

enzymes were published by Kasche et al. in 1971/73 [396, 387]. In the case of immobilized aerobic microorganisms and cells, the rate-determining process is oxygen-transfer into the spherical gel particles. The local oxygen level influences the viability of microorganisms and cells considerably [394, 398–400]. The effect of immobilization on the physiology of the cells has been investigated by several researchers, notably by Doran and Bailey on *Saccharomyces cerevisiae* [401]. Plasmid stability in immobilized microorganisms was investigated as well. Plasmid stability is improved by immobilization [402, 403]. Animal and plant cells were immobilized as well. They are discussed in 4.3.

4.2 High Density Cultures of Microorganisms

There is a general tendency to work with high-density cultures of microorganisms, in order to increase the volumetric productivity of reactors. High density of microorganisms can be obtained by:

- immobilization of microorganisms;
- application of hollow fibre reactors in continuous cultivation;
- reactors with cell recycling or cell membrane retention in continuous cultivation; or
- fed-batch cultivation.

The immobilization is treated in Sect. 4.1. Hollow fibre reactors are mainly used for high-density cultivation of animal cells. The main aim of their application for primary metabolite production by microorganisms is product removal during the cultivation. The hollow-fiber reactors were first applied in 1976/77 [404, 405]. The continuous cell recycle reactor, equipped with a centrifuge, had already been applied in 1960s [406, 407]. Margaritis and Wilke used a rotating-membrane filter in 1978 [408]; Rogers used a microchannel-membrane filter in 1980 [409]; and Nishizawa used a cross-flow-membrane module for ethanol production in 1983 [410]. Chang and Furusaki presented a review on early applications of membrane bioreactors [411].

Volumetric yields of lactic acid produced by *Lactobacilli* with immobilized microorganisms, in hollow-membrane, and in cell-recycling membrane reactors, were compared. These investigations indicate that the highest volumetric productivities were obtained in continuous cultures with immobilized cells and cell recycling [158]. The low productivity in hollow fiber reactors is due to the high pH-gradient in the reactor.

The highest cell concentrations were obtained in fed-batch operation. Bauer et al. had already obtained high cell densities in 1976 [412]. Still higher cell densities were obtained by suitable medium-composition and feeding strategies [413–422]. Korz et al. obtained 148 g DCW l^{-1} in a standard stirred-tank reactor [422] and Märkl obtained 174 g DCW l^{-1} in a dialysis reactor [423].

4.3 Animal and Plant Cell Cultures

Animal cell culture was developed in medicine towards the end of the 19th century [424–427]. However, modern industrial cell-culture began in the mid-1950s, with the use of animal cells for the research and development of vaccines. Standard cultivation media and cell cultivation techniques were developed in the period 1950–1965 [428, 429]. Propagation of human diploid cells on titanium disks began in 1969 [430]. van Wezel developed microcarriers for primary cells and cell strains in 1967 [431]. At first, these processes employed primary cells; later, normal cells – maintained as cell strains – were approved; at present, transformed cells, which have infinite life spans, are being used. However, the latter are not applied to the production of vaccines and materials for human use. For vaccine production, BHK cells [432] and other cells [433] are used. The development of animal-cell cultures is followed in the reports of the ESACT meetings [434–437]. Reviews of the development of large-scale cultivation of animal cells have been published by Spier [438, 439] and by Kelley et al. [440]. Special reviews of the early investigations of nerve and muscle, human hepatoma, lymphoid cells, and vaccine production, as well as of the production of tPA, are presented in Vol. 34,and on the production of MAB by hybridoma cells and of ß-interferon in Vol. 37 of Adv Biochem Eng [441, 442]. Special reviews on early investigations of plant tissue cultures are presented in Vols. 16 and 18 [443, 444] of the same series.

A list of genetically engineered vaccines is given by Ratafia [445].

Small reactors, low specific power input, and membrane aeration are often applied for animal tissue cultures. Therefore, no large fluid-dynamic stress effects are expected. By application of larger ST, BC and ATL reactors and direct aeration, gas-liquid interfacial effects and at high specific power-input the dynamic pressure of turbulence can impair the behaviour of animal cells, because of their sensitive plasma membrane. The mechanisms of gas and liquid dispersion in turbulent field was investigated by several researcher. The dynamic equilibrium bubble and droplet sizes depend on the eddy size or on the Kolmogorov-length-scale of turbulence, $\eta_K = \left(\frac{\nu}{E}\right)$ where η represents the viscosity and E is the energy dissipated per unit mass. Bubbles and drops larger than this length are unstable and are dispersed. The same holds true for the viability of the cells. Cells on microcarriers, flocs, pellets, and filamentous mycelia are usually damaged, because they are larger than η_k at high specific power-input. However, suspended animal cells are not damaged in directly aerated systems, because of their small size and the low specific power input, i.e. large η_k [446]. In aerated systems, bursting gas bubbles can destroy the cells [447, 448]. Animal cells are harmed by shear forces in pumps and valves, but are less so in indirectly aerated reactors. The effect of turbulent jet-flow on suspension cultures of plant cells was investigated by MacLoughlin et al. [449]. because of the selection pressure, the sensitivity of animal and plant cells against fluid-dynamic stresses gradually diminishes with time.

Anchorage-dependent animal cells grow on microcarriers. Non-adherent cells were first immobilized by Nilsson and Mosbach in 1979 [450] and by Nilsson et al. in 1983 [451]. Plant cells were first entrapped by Brodelius and Nilsson in 1980 [452] and subsequently by Brodelius and Mosbach [453]. Immobilized microorganisms and animal and plant cells were reviewed by Bierbaum et al. [454]. Hollow-fiber and microcapsule bioreactors are currently used for animal cell immobilization as well [455]. The encapsulation of living cells was first accomplished by Lim and Moss in 1981 [456]. Jarvis and Grdina developed a modified version of encapsulation (Encapel) in 1983 [457]. The successful use of hollow-fiber reactors for the cultivation of mammalian cells was first reported by Knazek et al. in 1972/74 [458, 459]. Since then, several other research groups have used hollow fiber systems for mammalian cells. A review of early investigations with hollow-fiber-immobilized mammalian cells was given by Heath and Belfort [455]. Recent results are published in the reports of ESAC meetings [460].

5
Conclusion

The combination of chemical-reaction engineering, biochemistry, and microbiology provide a good platform for the development of industrial production processes. The stoichiometry, the mass and energy balances, and the kinetics of chemical reactions were adapted to biochemical reactions. The standard techniques of chemical process monitoring and control were extended by highly specific enzymatic analyses and control strategies to batch and fed-batch cultivation processes. Process monitoring and control were adapted to immobilized-cell and high-cell-density cultures, as well as to animal and plant cell cultivations. Mathematical models were developed for the complex interrelation between the physiology of the cells and their physical and chemical environment. The discovery of the metabolic pathways allowed deeper insight into cell regulation, and thus to the development of structured mathematical models. Systematic investigations with advanced techniques under reproducible conditions permitted the validation of these mathematical models. Computer-aided process control and optimization became possible by means of well-established models. Reliable mass balances due to advanced process monitoring, intracellular in vitro enzyme activity, and metabolite concentration measurements, as well as in-vivo-NMR spectroscopy – in addition to the knowledge of the metabolic pathways – made metabolic flux analysis and metabolic engineering possible. Bioreaction engineering has developed during the last fifty years into bioreaction engineering science.

References

1. Russell I, Stewart GG (1995) Brewing. In: Rehm HJ, Reed G, Pühler A, Stadler P (eds) Biotechnology, 2nd completely rev'd edn. Reed G, Nagodawithana TW (eds) Vol 9 Enzymes biomass food and feed. VCH, Weinheim, p 419

2a. Pasteur L (1860) Ann Chim Phys 58:323

2b. Pasteur L (1862) Ann Chim Phys 64:5

2c. Pasteur L (1876) Etudes sur la bière. Paris 251, 271

3. Buchner E (1897) Ber dtsch Chem Ges 30:1174

4. Roehr M, Kubicek C, Kominek J (1995) In: Rehm HJ, Reed G, Pühler A, Stadler P (eds.) Biotechnology. 2nd completely Rev'd edn, Vol.6. Roehr M (ed) Products of primary metabolism. VCH, Weinheim, p 307

5. Bartolomew W, Karow EO, Sfat MR (1950) Ind Eng Chem 42:1827

6. Steel R, Maxon WD (1961) Ind Eng Chem 53:739

7. Lyons EJ (1970) Chem Eng Progr Symp Ser 66: No 100 p. 33

8a. Deindorfer FH. Humphrey AE (1959) Ind Eng Chem 51:809

8b. Deindorfer FH, West JM (1960) J Biochem Microbiol Technol Eng 2:165

9. Van der Beek CP, Roels JA (1984) Antonie van Leeuwenhoek 50:625

10a. Lengyel ZI, Nyiri L (1965) Biotechol Bioeng 7:91

10b. Lengyel ZI, Nyiri L (1966) Biotechnol Bioeng 8:337

11. Gaden EL, Humphrey AE (1959) Biochem Microbiol Techn Eng 1:413

12. Aiba S, Humphrey AE, Millis N (1964) Biochemical engineering, 1st edn. Academic, New York

13. Bailey JE, Ollis DF (1977) Biochemical engineering fundamentals. Mc Graw Hill, New York

14. Wang DIC, Cooney CL, Demain AL, Dunnill P, Humphrey AE, Lilly DL (1978) Fermentation and enzyme technology. Wiley, New York
15. Pirt SJ (1975) Principles of Microbe and Cell Cultivation. Blackwell, Oxford
16a. Moser A (1981) Bioprozesstechnik. Berechnungsgrundlagen der Reaktionstechnik biokatalytischer Prozesse. Springer, Wien
16b. Moser A (1988) Bioprocess Technology. Springer, New York
17. Bergter F (1983) Wachstum von Mikroorganismen, 2 Aufl. Chemie, Weinheim
18a. Schügerl K (1985) Bioreaktionstechnik, Band 1. O Salle Verlag, Frankfurt and Verlag Sauerländer, Aarau
18b. Schügerl K (1985) Bioreaction Engineering, Vol.1. Wiley, Chichester
19. Einsele A, Finn RK, Samhaber W (1985) Mikrobiologische und biochemische Verfahrenstechnik. VCH, Weinheim
20. Ward OP (1989) Fermentation Biotechnology. Wiley, Chichester
21a. Schügerl K (1991) Bioreaktionstechnik, Bd 2. O Salle Verlag, Frankfurt and Verlag Sauerländer, Aarau
21b. Schügerl K (1991) Bioreaction Engineering, Vol. 2. Wiley, Chichester
22. van't Riet K, Tramper J (1991) Basic bioreactor design. Marcel Dekker New York
23. Chmiel H (ed) (1991) Bioprozeßtechnik, Bd 1. Gustav Fischer Verlag, Stuttgart
24. Weide H, Páca J, Knorre WA (1991) Biotechnologie, 2 Aufl. Gustav Fischer, Jena
25. Nielsen J, Villadsen J (1994) Bioreaction engineering principles. Plenum, New York London
26. Blanch HW, Clark DS (1997) Biochemical Engineering. Marcel Dekker, New York
27. Büche W (1937) VDI-Zeitschr 37:1065
28a. Hixson AW, Baum SJ (1941) Ind Eng Chem 33:478, 1433
28b. Hixson AW, Baum SJ (1942) Ind Eng Chem 34:120, 194
28c. Hixson AW, Gaden Jr EL (1950) Ind Eng Chem 42:1792
29a. Kirillov NI (1940) J Appl Chem (USSR) 13:978
29b. Kirrilov NI (1945) J Appl Chem (USSR) 18:381
30. Bartholomew WH, Karow EO, Sfat MR, Wilhelm RH (1959) Ind Eng Chem 42:1801
31a. Cooper CM, Fernstrom GA, Miller SA (1945) Ing Eng Chem 36:504
31b. Rushton JH, Costich EW, Everett HJ (1950) Chem Eng Progr 46:395
31c. Rushton JH, Costich EW, Everett HJ (1951) Chem Eng Progr 47:467
31d. Rushton JH (1951) Chem Eng Progr 47:485
31e. Oldshue JY (1987) Personal communication
32a. Mason DR, Piret EL (1950) Ind Eng Chem 42:817
32b. MacDonald RW, Piret EL (1951) Chem Eng Progr 47:363
33a. Metzner AB, Otto RE (1957) AIChE Journal 3:3
33b. Metzner AB, Feehs RH, Lopez Ramos H, Otto RE, Tuthill JD (1961) AIChE J 7:3
34. Aiba S (1958) AIChE J 4:485
35. Yoshida F, Miura Y (1960) Ind Eng Chem 52:435
36. van Krevelen PW, Hoftijzer PJ (1948) Rec Trav Chim 67:563
37a. Kramers H Chem Eng Sci (1958) 8:41
37b. Hanhart J, Kramers H, Westerterp KR (1963) Chem Eng Sci 18:503
38a. Calderbank PH (1958) Trans Inst Chem Eng 36:443
38b. Calderbank PH (1959) Trans Inst Chem Eng 37:173
38c. Calderbank PH. Moo-Young MB (1959) Trans Inst Chem Eng 37:26
39. Deckwer WD (1991) Bubble Column Reactors. Wiley, Chichester
40. Chisti Y (1989) Airlift Bioreactors. Elsevier Sci, Amsterdam
41. Patanakar SV(1980) Numerical Heat Transfer and Fluid Flow. McGraw-Hill, New York
42. Fokema MD, Kresta SM (1994) Can J Chem Eng 72:177
43. Abid M, Xuereb C, Bertrand J (1994) Can J Chem Eng 72:184
44. Hjertager BH (1996) Computational fluid dynamics (CFD) modelling and simulation of bioreactors. In: Measuring and control, Course notes of Bioreactor engineering. EFB-WP and Bioreactor performance EFB-WP
45. Jenne M, Reuss M (1996) Chem Ing Techn 68:295

46. Lapin A, Lübbert A (1994) Chem Eng Sci 49:3661
47. Becker S, Sokolichin A, Eigenberger G (1994) Chem Eng Sci 49:5747
48. Sokolichin A, Eigenberger G (1994) Chem Eng Sci 49:5735
49. Lübbert A, Paaschen T, Lapin A (1996) Biotechnol Bioeng 52:248
50. Sokolichin A, Eigenberger G, Lapin A, Lübbert A (1997) Chem Eng Sci 52:611
51. Delnoij E, Kuipers JAM, van Swaaij WPM (1997) Chem Eng Sci 52:3623
52a. Minkevich G, Eroshin VK (1973) Folia Microbiol 18:376
52b. Minkevich G, Eroshin VK (1974) Microbiol. Promishlen (in Russian) 82:103
52a. Erickson LE, Minkevich IG, Eroshin VK (1978) Biotechnol Bioeng 20:1595
53b. Erickson LE (1980) Biotechnol Bioeng 22:1929
54a. Roels JA (1980) Biotechnol Bioeng 22:33
54b. Roels JA (1980) Biotechnol Bioeng 22:2457
54c. Roels JA (1983) Energetics and kinetics in biotechnology. Elsevier Biomed, Amsterdam
55a. Stouthamer AH (1973) Antonie van Leeuwenhoek. J Microbiol Serol 39:545
55b. de Kwaadsteniet JW, Jager JC, Stouthamer AH (1976) J Theor Biol 57:103
56. Yamané T, Shitani T (1981) Biotechnol Bioeng 23:1373
57. Peringer P, Blachere H, Corrieu G, Lane AG (1947) Biotechnol Bioeng 26:431
58. Zabriskie DW, Humphrey AE (1976) AIChE J 24:138
59. Cooney CL, Wang HY, Wang DIC (1977) Biotechnol Bioeng 19:55
60. Wang HY, Cooney CL, Wang DIC (1977) Biotechnol Bioeng 19:69
61. Geurts ThGE, de Kok HE, Roels JA (1980) Biotechnol Bioeng 22:2031
62. Dekkers JGJ, de Kok HE, Roels JA (1981) Biotechnol Bioeng 23:1023
63. San KY, Stephanopoulos G (1984) Biotechnol Bioeng 26:1209
64. Heijnen SJ, Roels AJ, Stouthamer AH (1979) Biotechnol Bioeng 21:2175
65. Mou DG, Cooney CL (1983) Biotechnol Bioeng 25:225
66. Solomon BO, Erickson LE, Yang SS (1983) Biotechnol Bioeng 25:2683
67. Grosz R, Stephanopoulos G (1983) Biotechnol Bioeng 25:2149
68. San KY, Stephanopoulos G (1984) Bioetchnol Bioeng 26:1189
69. Grosz R, Stephanopoulos G, San KY (1984) Biotechnol Bioeng 26:1198
70. Lee HY, Erickson LE, Yang SS (1985) Biotechnol Bioeng 27:1640
71. Ross A, Schügerl K (1987) In: Neissel OM, van der Meer RR, Luyben KChAM (eds) Proc 4th Eur Congr Biotechnol, Vol. 3. Elsevier Sci, Amsterdam, p 94
72. Reuss M, Fröhlich S, Kramer B, Messerschmidt K, Niebeschütz H (1984) 3rd Eur Congr on Biotechnol, Vol. 2. Verlag Chemie, Weinheim, p 455
73. Heinzle E, Dettweiler B (1986) Ann Biotechnol Congr DECHEMA
74. Esener AA, Veerman T, Roels JA, Kossen NW (1982) Biotechnol Bioeng 24:1749
75. Jansen NB, Flickinger MC, Tsao GT (1984) Biotechnol Bioeng 26:573
76. Zeng AP, Ross A, Deckwer WD (1990) Biotechnol Bioeng 36:965
77. Solomon BO, Oner MD, Erickson LE Yang SS (1984) AIChE J 30:747
78. Papoutsakis E T, Meyer CL (1985) Biotechnol Bioeng 27:67
79. Solomon BO, Erickson LE (1981) Process Biochem, Febr March: 44
80. Heijnen JJ, Dijken JP (1992) Biotechnol Bioeng 39:833
81. Hinshelwood CM (1946) Chemical Kinetics of the Bacterial cell, Oxford Univ Press, London
82. Michaelis L, Menten ML (1913) Biochem Z 49:333
83. Laidler KJ (1958) The chemical kinetics of enzyme action, Oxford Univ Press,
84. Monod J (1949) Ann Rev Microbiol 3:371
85. Gaden EL (1955) Chem & Ind 154
86. Contois D (1959) J Gen Microbiol 21:40
87. Luedeking R, Piret EL (1959) J Biochem Microbiol Tech & Eng 1:393
88. Deindorfer FH (1960). In: Umbreit WW (ed) Fermentation kinetics and model process. Adv Appl Microbiol, Vol. 2 Academic, New York, 2:321
89a. Kono T (1968) Biotechnol Bioeng 10:105
89b. Kono T, Asai T (1969) Biotechnol Bioeng 11:293
90. Tsao GT, Yang CM (1976) Biotechnol Bioeng 23:1827

91. Levenspiel O (1980) Biotechnol Bioeng 22:1671
92. Monod J (1950) Ann Inst Pasteur 79:390
93a. Novick A, Szilárd L (1950) Science 112:715
93b. Novick A, Szilárd L (1959) Proc Acad Sci 36:708
94. Spicer CC (1950) Biometrics 11:225
95. Herbert D, Elsworth R, Telling RC (1956) J Gen Microbiol 14:601
96. Moser H (1957) Proc Acad Sci 43:222
97a. Malek I (ed) (1958) Continuous cultivation of microorganisms, 1st Symp Publ House of Czech Acad Sci, Prague
97b. Malek I (ed) (1964) Continuous cultivation of micro organisms 2nd Symp Publ House of Czech Acad Sci, Prague
98. Luedenking R, Piret E (1959) J of Biochem Microbiol Tech &. Eng 1:431
99. Gerhardt P, Bartlett MC (1959) Continuous industrial fermentation. In: Umbreit WW (ed) Adv in Appl Microbiol. Academic, New York, 1:215
100. Koga S, Humphrey AE (1967) Biotechnol Bioeng 9:375
101. Yano T, Koga S (1969) Biotechnol Bioeng 11:139
102. Edwards VH, Ko RC, Balogh SA (1972) Biotechnol Bioeng 14:939
103a. Meyerhof O, Kiesling F (1933) Biochem Z 264:62
103b. Meyerhof O, Kiesling F (1934) Biochem Z 267:345
103c. Meyerhof O, Kiesling F (1935) Biochem Z 276:252; 279:40
104a. Warburg O, Christian W (1941) Naturwiss, 589
104b. Warburg O, Christian W (1943) Biochem Z 314:149
105. Lipmann F (1941) Metabolic generation and utilization of phosphate bond energy. Adv Enzymol 18:99
106. Warburg O (1948) Wasserstoffübetragende Fermente. Saenger, Berlin
107. Krebs HA, Kornberg HL (1957) Energy transformation in living matter. Springer, Berlin
108. Axelrod B (1967) Glycolysis. In: Greenberg DM (ed) Metabolic Pathways, 3 edn, Vol 1. Academic, New York, p 112
109. Doelle HW (1975) Bacterial Metabolism. Academic, New York
110. Gottschalk G (1985) Bacterial Metabolism 2nd edn. Springer, New York
111. Clark LC (1956) Am Soc Artif Organs 2:41
112. MacInnes DA, Dole M (1930) J Am Chem Soc 52:29
113. Perley GA (1945) British Patent 574 029
114. Scheller F, Schubert F (1989) Biosensoren. Akademie, Berlin
115. Free AH, Adams EC, Kercher ML, Free HM, Cook MHM Abstr Int Congr Clinical Chemistry, New York, p 235
116. Clark LC, Lyons C (1962) Ann NY Acad Sci 102:29
117. Updike SJ, Hicks, GP (1967) Nature 214:986
118. Reitnauer PG (1972) DDR Patent 191 229
119. Yellow Springs Laboratory (1970) Yellow Springs, Ohio, USA
120. Rancine P, Klenk HO, Kochsiek K (1975) Z Klin Chem Klin Biochem 13:533
121. Mosbach K (1977) US Patent 4021 307
122. Loewe CR, Goldfinch MJ (1983) Biochem Soc Trans 11:446
123. Divies C (1975) Ann Microbiol (Paris) 126:175
124. Guibault GG (1976) Handbook of Enzymatic Methods of Analysis. Marcel Dekker, New York, p 490
125. Ho MYK, Rechnitz GA (1987) Anal Chem 59:536
126. Janata J (1975) J Am Chem Soc 97:2914
127. Belli SL, Rechnitz GA (1986) Anal Lett 19:403
128. Turner APF, Heinemann WR, Karube I, Schmid RD (eds) (1992) Biosensors'92. Proc 2nd Word Congress on Biosensors, Geneva. Elsevier Advanced Technol, Oxford, UK
129. Schmid RD (ed) (1990) Flow injection analysis (FIA) based on enzymes and antibodies. VCH, Weinheim
130. Schügerl K (1997) Bioreaction Engineering, vol 3. Bioprocess Monitoring. Wiley, Chichester

131. Ruzicka J, Hansen EH (1975) Anal Chim Acta 106:207
132. Ruzicka J, Hansen EH (1998) Flow Injection analysis. 2nd edn. Wiley, New York
133. Schmid RD, Künnecke W (1990) J Biotechnol 14:3
134. Leisola M, Kauppinnen V (1978) Biotechnol Bioeng 20:837
135. Leisola M, Virkunnen J, Karvonen E, Meskanen A (1979) Enzyme Mircrob Technol 1:117
136. Seifert GK, Matteau PP (1988) Biotechnol Bioeng 32:923
137. Appelquist R, Johannson G, Holst O, Mattiasson B (1989) Anal Chim Acta 216:299
138. Ferrari A, Gerke JR, Watson RW, Umbreit WW (1965) Ann NY Acad Sci 130:704
139. Gerke JR (1965) Ann NY Acad Sci 130:722
140. Zabriskie DW, Humphrey AE (1978) Eur J Appl Microbiol Biotechnol 35:337
141. Recktenwald A, Kroner KH, Kula MR (1985) Enzyme Microb Technol 7:607
142. Garn M, Gisin M, Thommen C, Cevey P (1989) Bioetchnol Bioeng 43:234
143. Graf H, Wentz D, Schügerl K (1991) Biotechnol Techn 5:183
144. Holst O, Hakanson H, Miyabashi A, Mattiasson B (1988) Appl Microbiol Biotechnol 28:32
145. Schmidt W, Meyer HD, Schügerl K, Kuhlmann W, Bellgardt KH (1984) Anal Chim Acta 163:101
146. Brandt J, Hitzmann B (1994) Anal Chim Acta 291:29
147. ANASYSCON GmbH, Hannover
148. BIOSPECTRA, Schlieren (Zürich)
149. TECATOR, Rodgau
150. Spruytenberg R, Dunn IJ, Bourne JR (1979) Biotechnol Bioeng Symp 9:359
151. Pons MN, Engasser JM (1988) Anal Chim Acta 213:231
152. Filippini C, Moser JU, Sonnleitner B, Fiechter A (1991) Anal Chim Acta 255:91
153. Dincer AK, Kalyanpur M, Skea W, Ryan M, Kiersted T (1984) Continuous on-line monitoring of fermentation processes, In: Developments in Industrial Microbiology, vol. 25. Soc of Ind Microbiology
154. Bayer T, Herold T, Hiddessen R, Schügerl K (1986) Anal Chim Acta 190:213
155. Heinzle E, Dunn, IJ (1991) Methods and Instruments in Fermentation Gas Analysis. In: Schügerl K (ed) Measuring Modelling and Control. vol 4. In: Rehm HJ, Reed G, Pühler A, Stadler P (eds) Biotechnology, 2nd edn. VCH, Weinheim, p 27
156. Hammond SV, Brookes IK (1992) In: Ladish M, Bose A (eds) Harassing Biotechnology for the 21st Century. ACS, Washington, p 325
157. Tartakowski B, Sheintuch M, Hilmer JM, Scheper T (1996) Biotechnol Progr 12:126
158. Schügerl K (1991) On-line analysis of broth. In: Schügerl K (ed) Measuring, Modelling and Control, Vol 4. In: Rehm HJ, Reed G, Pühler A, Stadler P (eds) Biotechnology. 2nd edit. VCH, Weinheim, p 149
159. Douglas JM (1972) Process Dynamics and Control, Prentice Hall, Englewood Cliffs,
160. Koppel LB (1968) Introduction to Control Theory, Prentice Hall, Englewood Cliffs
161. Armiger WB, Humphrey AE (1979) In: Peppler HJ, Perlman D (eds) Microbial Technology, vol 2. Academic, New York, p 345
162. Bull DN (1977) Annu Rep Ferment Processes 6:359
163. Hampel WA (1979) Adv Biochem Eng 13:1
164. Hatch RT (1982) Annu Rep Ferment Proc 5:291
165. Jefferis RP (1975) Process Biochem 10:15
166. Rolf MJ, Lim HC (1982) Enzym Microb Technol 4:370
167. Weigand WA, (1978) Annu Rep Ferment Processes 2:43
168a. Zabriskie DW, Armiger WB, Humphrey AE (1977) In: Workshop Computer Applications in Fermentation Technology GBF Monogr Ser 3:59
168b. Zabriskie DW (1979) Ann NY Acad Sci 326:223
169. Halme A (ed) (1982) Modelling and control of biotechnical processes proceed workshop, Pergamon Press, Oxford
170. Hermann JPR (eds) (1982) Computer Application in Fermentation Technology Soc Chem Ind, London
171. Johnson A (ed) (1985) 1st IFAC Symposium on Modelling and Control of. Biotechnological Processes, Pergamon Press, Oxford

172. Stephanopoulos G, Karim N (eds) (1992) 2nd IFAC Symposium on Modelling and Control of Biotechnical Processes
173. Munack A, Schügerl K (eds) (1995) Computer Applications in Biotechnology 6th Internat. Conference. Elsevier Sci, Amsterdam
174. Wang NS, Stephanopoulos GN (1984) Computer application to fermentation processes, Critical Reviews in Biotechnology, vol 2 Issue 1. CRC Press, p 1
175. Lim H, Lee KS (1991) Control of Bioreactor Systems. In: Schügerl K (ed) Measuring modelling and montrol, vol 4. In: Rehm HJ, Reed G (eds) Biotechnology. VCH, Weinheim, p 509
176. Bastin G, Dochain D (1990) Online estimation and Adaptive Control of Bioreactors, Elsevier Sci Publ, Amsterdam
177. Blackmann FF (1905) Ann Bot 19:281
178. Monod J (1942) Recherches sur la Crossiance des Cultures Bacteriennes, Paris, Hermann & Cie
179. Teissier G (1936) Ann Physiol Physicochem Biol 12:527
180. Moser H (1958) The Dynamics of Bacterial Populations Maintained in the Chemostat, Carnegie Institution, Washington DC Publ No 614
181. Herbert D (1958) Continuous culture of micro-organisms. Some theoretical aspects. In: Malik I (ed) Continuous cultivation of micro-organisms. Czech Acad Sci Prague
182. Kargi F (1977) J App Chem Biotechnol 27:704
183. Bellgardt KH (1991) Cell models. In: Schügerl K (ed) Measuring, Modelling and Control Vol 4 In: Rehm HJ, Reed G (eds) "Biotechnology", 2nd edn. VCH, Weinheim, p 276 (Table 4)
184. Tsuchiya HM, Frederickson AG, Aris R (1966) Adv Chem Eng 6:125
185. Dhurjati P, Ramkrishna D, Flickinger MC, Tsao GTA (1985) Biotechnol Bioeng 27:1
186. Kompala D, Jansen N, Tsao G, Ramkrishna D (1986) Biotechnol Bioeng 28:1044
187. Domach MM, Shuler ML (1984) Biotechnol Bioeng 26:877
188. Domach MM, Leung SK, Cahn RE, Cocks GG, Shuler ML (1984) Biotechnol Bioeng 26:203
189. Peretti SW, Bailey JE (1986) Biotechnol Bioeng 28:1672
190. Peretti SW, Bailey JE, Lee JJ (1989) Biotechnol Bioeng 34:902
191. Hatzimanikatis V, Lee KH, Renner WA, Bailey JE (1995) Biotechnol Lett 17:668
192. Bailey JE (1991) Science, 252:1668
193. Harder A, Roels JA (1982) Adv Biochem Eng 21:55
194. Jacob F, Monod J (1961) J Mol Biol 3:318
195. Bellgardt KH, Yuan J (1991) Process Models. Optimization of yeast production. A case study. In: Schügerl K (ed) Measuring, Modelling and Control, vol 4. In: Rehm HJ, Reed G, Pühler A, Stadler P (eds) Biotechnology, 2nd edn. VCH, Weinheim, p 383
196. Steel R, Maxon WD (1961) Ind Eng Chem 9:739
197. Steel R, Maxon WD (1962) Biotechnol Bioeng 4:231
198. Steel R, Maxon WD (1966) Biotechnol Bioeng 8:97, 109
199. Cooney CL, Wang DIC (1971) Biotechnol Bioeng Symp No 2:63
200. Buckland BC, Gbewonyo K, DiMasi D, Hunt G, Westerfield G, Nienow AW (1988) Biotechnol Bioeng 31:737
201. Nienow AW (1990) TIBTECH 8:224
202. Schügerl K (1990) J Biotechnol 13:251
203. Pons A, Dussap CG, Gross JB (1987) In: Neissel OM, van der Meer RR, Luyben KChAM (eds) Proc 4th Europ Congr Biotechnol, 1, Elsevier Sci, Amsterdam
204. Suh IS, Schumpe A, Deckwer WD (1992) Biotechnol Bioeng 39:85
205. Gebauer A, Scheper T, Schügerl K (1987) Bioproc Eng 2:13
206. Adler I, Fiechter A, (1983) Chem Ing Techn 55:322
207. Adler I, Fiechter A (1986) Bioproc Eng 1:51
208. Peters HU, Suh IS, Schumpe A, Deckwer WD (1992) Can J Chem Eng 70:742
209a. Herbst H, Peters HU, Suh IS, Schumpe A, Deckwer WD (1988) Chem Ing Techn 60:407
209b. Herbst H, Schumpe A, Deckwer WD (1992) Chem Eng Techn 15:425

210. Yonsel S, Deckwer WD0. (1990) Bio Eng 6:12
211. Schlüter V, Yonsel S, Deckwer WD (1992) Chem Ing Techn 64:474
212. Funhashi H, Maehera M, Taguchi T, Yoshida T (1987) J Chem Eng Japan 20:16
213. Schügerl K (1982) Adv Biochem Eng 22:93
214. Lübbert A, Larson B, Wan LW, Bröring S (1990) ICHME E Symp Ser No 121:203
215. Shamlou PA, Pollard DJ, Isou AP, Lilly MD (1994) Chem Eng Sci 49:303
216. Lübbert A, Fröhlich S, Schügerl K (1987) In: Heinzle E, Reuss M (eds) Mass spectrometry in biotechnol process analysis and control. Plenum, New York, p 125
217. Rüffer HM, Wan Liwei, Lübbert A, Schügerl K (1994) Bioproc Eng 11:153
218a. Lotz M, Fröhlich S, Matthes R, Schügerl K, Seekamp M (1991) Process Biochem 26:301
218b. Fröhlich S, Lotz M, Korte T, Lübbert A, Schügerl K, Seekamp M Biotechnol Bioeng 37:910
218c. Fröhlich S, Lotz M, Larson B, Lübbert A, Schügerl K, Seekamp M Biotechnol Bioeng 38:56
219. Mayr B, Nagy E, Horvat P, Moser A (1993) Chem Biochem Eng 7:31
220. Oosterhuis NMG, Kossen NWF (1984) Biotechnol Bioeng 26:546
221. Oosterhuis NMG (1984) Dissertation, TU Delft
222. Fields PR, Slater NKH (1984) Biotechnol Bioeng 26:719
223. Stravs AA, Pittet A, von Stockar U, Reilly PJ (1986) Biotechnol Bioeng 28:1302
224. Pedersen AG, Bundgaard-Nielsen M, Nielsen J (1994) Bioetchnol Bioeng 44:1013
225. Rüffer HM, Pethoe A, Schügerl K, Lübbert A, Ross A, Deckwer WD Bioproc Eng 11:145
226a. Reuss M, Bajpai RK, Berke W (1982) J Chem Technol Biotechnol 32:81
226b. Bajpai RK, Reuss M (1982) Can J Chem Eng 60:384
227. Moes J, Griot M, Keller J, Heinzle E, Dunn IJ, Bourne JR (1985) Biotechnol Bioeng 27:482
228. Kawase Y, Moo-Young M (1989) Appl Microbiol Biotechnol 30:596
229. Moo-Young M, Halard B, Allen DG, Burrell R, Kawase Y (1987) Biotechnol Bioeng 30:746
230. Kawase Y, Halard B, Moo-Young M (1992) Biotechnol Bioeng 39:1133
231. Pigache S, Trystram G, Dhoms P (1992) Biotechnol Bioeng 39:923
232a. Buchholz H, Luttmann R, Zakrzewski W, Schügerl K (1981) Europ Appl Microbiol Biotechnol 12:63
232b. Luttmann R, Munack A, Thoma M (1985) Adv Biochem Eng 32:95
232c. Isaacs S, Thoma M, Munack A (1987) Biotechnol Progr 3:248
233. Manfredini R, Cavallera V, Marini L, Donati G (1983) Biotechnol Bioeng 25:3115
234. Radez T, Hudcova V, Koloini T (1991) Chem Eng Journal 46:B83
235. Carrington R, Dixon K, Harrop AJ, Macaloney G (1992) In Ladisch M, Bose A (eds.) Harnessing biotechnol. for the 21st century, p. 183
236. Jurecic R, Berovic M, Steiner W, Koloini T (1984) Can J Chem Eng 62:334
237. Taguchi H, Miyamoto S (1966) Biotechnol Bioeng 8:43
238. Rho D, Mulchandani JH, Luong JHT, LeDuy A (1988) Appl Microbiol Biotechnol 28:361
239. Schügerl K, Lücke J, Lehmann J, Wagner F (1978) Adv Biochem Eng 8:63
240. Rüffer HM, Bröring S, Schügerl K (1995) Bioproc Eng 12:119
241. Young MA, Carbonell RG, Ollis DF (1991) AIChE Journal 37:403
242. Heinzle E, Moes J, Dunn IJ, (1983) Biotechn Lett 7:235
243. Reuss M, Brammer U (1985) In: Johnson A (ed) Modelling and pontrol of piotechnol processes" 3 IFAC Symp Ser Pergamon Press, Oxford, p. 119
244. Sweere APJ, Luyben KChAM, Kossen NWF (1987) Enzyme Microb Technol 9:386
245. Sweere APJ, Mesters JR, Janse L, Luyben KChAM, Kossen NWF (1988) Biotechnol Bioeng 31:567
246. Sweere APJ, Janse L, Luyben KChAM, Kossen NWF (1988) Biotechnol Bioeng 31:579
247. Namdew PK, Yegneswaran PK, Thompson BG, Gray MR (1991) Can Chem Eng 69:513
248. Abel C, Hübner U, Schügerl K (1994) J Biotechnol 32:45
249. Abel C, Linz F, Scheper T, Schügerl K (1994) J Biotechnol 32:183
250. Schügerl K (1992) In: Ladish M, Bose A (eds) Harnessing biotechnol for the 21st century. Am Chem Soc, Washington, p 232
251. Schügerl K (1993) Bioprocess Eng 9:215

252. Liao J, Delgado J (1993) Biotechnol Progr 9:221
253. Heinrich R, Rapoport TAA (1974) Eur J Biochem 42:89
254. Higgins J (1965) In: Chance B, Estabrook RW, Williamson JR (eds) Control of Energy Metabolism. Academic, New Work, p 13
255. Kacser H, Burns JA (1973) In: Davies DD (ed) Rate Control of Biological Processes. Cambridge Univ Press, Cambridge, p 65
256. Savageau MA (1971) Nature 229:542
257. Luong JHT, Voleski B (1980), Can J Chem Eng 58:497
258. Owusu RK, Finch A (1985) J Gen Appl Microbiol 21:221
260. Marison IW, von Stockar U (1986) Biotechnol Bioeng 28:1780
261. Roy S, Samson R (1988) J Biotechnol 8:193
262. Larsson C, Blomberg A, Gustafsson L (1991) Biotechnol Bioeng 38:447
263. Meier-Schneiders M, Grosshans U, Busch C, Eigenberger G (1995) 43:431
264. Ataai MM, Shuler ML (1985) Biotechnol Bioeng 27:1027
265. de Koning W, van Dam K (1992) Anal Biochem 204:118
266a. Theobald U, Mailinger W, Reuss M, Rizzi M (1993) Anal. Biochem 214:31
266b. Theobald U, Mailinger W, Baltes M, Rizzi M, Reuss M (1997) Biotechnol Bioeng 55:305
267. Zupke C, Sinskey AJ, Stephanopoulos G (1995) Appl Microbiol Biotechnol 44:27
268. Nielsen J, Jorgensen HS (1995) Biotechnol Progr 11:299
269. Weuster-Botz D, de Graaf AA (1996) Adv Biochem Eng Biotechnol 54:75
270. Koplove HM, Cooney CL (1976) Anal Biochem 72:297
271. Ahlmann N, Niehoff A, Rinas U, Scheper T, Schügerl K (1986) Anal Chim Acta 190:221
272. Blankenstein G, Kula MR (1991) Anal Chim Acta 248:371
273. Kracke-Helm HA, Brandes L, Hitzmann B, Rinas U, Schügerl K (1991) J Biotechnol 20:95
274. Steube K, Spohn U (1994) Anal Chim Acta 287:235
275. Locher G, Hahnemann U, Sonnleitner B, Fiechter A (1993) J Biotechnol 29:57; 75
276. Rizzi M, Theobald U, Querfurth E, Rohrhirsch T, Baltes M, Reuss M (1996) Biotechnol Bioeng 49:316
277. Zeng AP, Deckwer WD (1994) J Biotechnol 37:67
278a. Stephanopoulos G, San KY (1984) Biotechnol Bioeng 26:1176
278b. San KY, Stephanopoulos G (1984) Biotechnol Bioeng 26:1209
279. Dochain D, Perrier M (1997) Adv Biochem Eng Biotechnol 56:147
280. Goel A, Ferrenace J, Jeong J, Ataai M (1993) Biotechnol Bioeng 42:686
281a. Sauer U, Hatzimanikatis V, Hohmann P, Mannberg M, van Loon APGM, Bailey JE (1996) Appl Environm Microbiol 50:3687
281b. Sauer U, Hatzimanikatis V, Bailey JE, Hochuli M, Szyperski T, Wütrich K (1997) Nature Biotechnol 15:448
282. Park SM, Shaw-Reid C, Sinskey AJ (1997) Appl Microbiol Biotechnol 47:430
283. Vallino JJ, Stephanopoulos G (1993) Biotechnol Bioeng 41:633
284. Marx A, de Graaf AA, Wiechert W, Eggeling L, Sahm H (1996) Biotechnol Bioeng 49:111
285. Takiguchi N, Shimizu H, Shioya S (1997) Biotechnol Bioeng 55:170
286. Wiechert W, Siefke C, de Graaf AA, Marx A (1997) Bioetchnol Bioeng 55:118
287. Pons A, Dussap CG, Réquignot C, Gros JB (1996) Biotechnol Bioeng 51:177
288. Ko WF, Bentley W, Weigand WA (1993) Biotechnol Bioeng 42:843
289. Reagan I, Gregory M (1995) J Biotechnol 42:151
290. Sharfstein S T, Tucker SN, Mancuso A, Blanch HW (1994) Biotechnol Bioeng 43:1059
291. Zupke C, Stephanopoulos G (1995) Biotechnol Bioeng 45:292
292. Zupke C, Stephanopoulos G (1994) Biotechnol Progr 10:498
293. de Noronha Pissara P, Nielsen J, Bazin MJ (1996) Biotechnol Bioeng 51:168
294. de Noronha Pissara J, Nielsen J (1997) Biotechnol Progr 13:156
295. Jorgensen H, Nielsen J, Villadsen J (1995) Biotechnol Bioeng 46:117
296. Galazzo JL, Bailey JE (1989) Biotechnol Bioeng 33:1283
297. Rizzi M, Theobald U, Querfurth E, Rohrhirsch T, Baltes M, Reuss M (1996) Biotechnol Bioeng 49:316
298. Theobald U, Mailinger W, Baltes M, Rizzi M, Reuss M (1997) Biotechnol Bioeng 55:305

299. Rizzi M, Baltes M, Theobald U, Reuss M (1997) Biotechnol Bioeng 55 : 592
300. Zeng AP, Deckwer WD (1994) J Biotechnol 37 : 67
301. Jin S, Ye K, Shimizu K (1997) J Biotechnol 54 : 161
302. Ferrance JP, Goel A, Ataai MM (1993) Biotechnol Bioeng 42 : 697
303. Venkatesh KV (1997) Proc Biochem 32 : 651
304. Snoep JL, Arfman N, Yomano LP, Westerhoff HV, Conway T, O'Neal Ingram L (1996) Biotechnol Bioeng 51 : 190
305. de Graaf AA, Schoberth SM, Probst U, Wittig RM, Strohhäcker J, Sahm H (1991) Chem Ing Techn 63 : 628
306. Wiechert W, de Graaf AA (1997) Biotechnol Bioeng 55 : 101
307. Stephanopoulos G (ed) (1998) Special issue on metabolic engineering. Biotechnol Bioeng 58 : 119–343
308. Barr A, Feigenbaum EA (1981) The Handbook of Artificial Intelligence, vol I. William Kaufmann, Los Altos
309. Winston PH (1984) Artificial Intelligence. 2nd ed. Addison Wesley, Massachusetts
310. Jackson P (1986), Introduction to Expert Systems. Addison-Weseley, Wokingham, UK
311. Halme A, Karim N (1991) Expert Systems for Biotechnology. In: Schügerl K (ed) Measuring, Modelling and Control. Vol 4 In: Rehm HJ, Reed G, Pühler A, Stadler P (eds) Biotechnology 2nd rev'd edn. VCH, Weinheim, p. 625
312. Russo MF, Peskin RL (1987) Knowledge-based systems for the engineer Chem Eng Progr 83 : 38
313. Stephanopoulos G (1987) Chem Eng Prog 83 : 44
314. Cooney CL, O'Connor G, Riera FS (1988) An expert eystem for intelligent supervisory control of fermentation. In: Durant G, Bobichon L, Florent J (eds) 8th Internat Biotechnol Symp Paris. Soc Franc de Microbiol, p 563
315. Linko P, Zhu YH (1992) In: Karim MN, Stephanopoulos G (eds) Modeling and control of botechnical pocesses. IFAC Symp Ser Pergamon Press, Oxford, p 163
316. Zhang XC, Visala A, Halme A, Linko P (1994) J of Biotechnol 37 : 1
317. Karim MN, Halme A (1988) Reconciliation of measurement data in fermentation using on-line expert system. In: Fish NM, Fox RI (eds.) Computer Application in Fermentation Technology: Modelling and Control of Biotechnical Processes. 4th Intern Congress. Ellis Horwood, Chichester.
318. Halme A (1988) Expert system approach to recognise the state of fermentation and diagnose faults in bioreactors. In: Proc 4th IntSymp on Computer Applications in Fermentation Technol, Cambridge, UK p 159
319a. Lübbert A, Hitzmann B (1987) Hung J Ind Chem 15 : 39
319b. Lübbert A, Hitzmann B, Kracke-Helm HA, Schügerl K (1988) On experiences with expert systems in the control of bioreactors. In: Proc 4th Int Symp on Computer Applications in Fermentation Technol, Cambridge, UK p 297
320. Shi Z.P, Shimizu K (1992) In: Karim MN, Stephanopoulos G (eds) Modeling and control of biotechnical processes. IFAC Symp Ser Pergamon Press, Oxford, p 167
321. Hitzmann B, Lübbert A, Schügerl K (1992) Biotechnol Bioeng 39 : 33
322. Konstantinov KB, Yoshida T (1992) J of Biotechnol 24 : 33
323. Konstantinov KB, Aarts R, Yoshida T (1993) Adv In Biochem Eng Biotechnol 48 : 169
324. Stephanopoulos G, Stephanopoulos GN (1986) Trends Biotechnol, Elsevier Sci Pub, Amsterdam, p 241
325. Bustamante ZR, Pokkinen M, Takuwa T, Asama H, Linko P, Endo I (1992) J Biotechnol 24 : 75
326. Nakajima M, Siimes T, Yada H, Asama H, Nagamune T, Linko P, Endo I, In: Karim MN, Stephanopoulos G (eds.) Modeling and control of biotechnical processes. IFAC Symp Ser Pergamon Press, Oxford, p 179
327. Rivera SL, Karim MN (1992) In: Karim MN, Stephanopoulos G (eds) Modeling and Control of Biotechnical Processes. IFAC Symp Ser Pergamon Press, Oxford, p 159
328. Aynsley M, Hofland A, Morris AJ, Montague GA, Di Massimo C (1993) Adv in Biochem Eng Biotechnol 48 : 1

329. Shuqing W, Wei Z (1992) In: Karim MN, Stephanopoulos G (eds) Modeling and control of biotechnical processes. IFAC Symp Ser Pergamon Press, Oxford, p 437
330. Qi C, Wang SQ, Wang JC (1988) Application of expert system to the operation and control of industrial antibiotic fermentation process. In: 4th Internat Congr on Computer Appl in Ferment Technol. Cambridge, UK
331. Horiuchi J, Kishimoto M (1995) In: Munack A, Schügerl K (eds) 6th Internat Conference on Computer Applications in Biotechnology. Pergamon Press, Oxford, p 47
332. Simutis R, Havlik I, Lübbert A (1992) J of Biotechnol 24:211
333. Simutis R, Havlik I, Lübbert A (1993) J of Biotechnol 27:203
334. Simutis R, Havlik I, Dors M, Lübbert A (1995) In: Munack A, Schügerl K (eds) 6th Internat Conference on Computer Applications in Biotechnology, Pergamon Press, Oxford, p 59
335. Aarts RJ, Suviranta A, Rauman-Aalto P, Linko P (1989) An expert system in enzyme production control, In: Proc Int Conf On Biotechnol and Food. Stuttgart, Germany, p 20
336. Dors M, Simutis R, Lübbert A (1995) In: Munack A, Schügerl K (eds) 6th Internat Conference on Computer Applications in Biotechnology. Pergamon Press, Oxford, p 72
337. Simutis R, Oliviera R, Manikowski M, de Azevedo F, Lübbert A (1997) J Biotechnol 59:73
338. Oh GS, Eikens B, Yoshida T, Karim MN (1995) In: Munack A, Schügerl K (eds) 6th Internat Conference on Computer Applications in Biotechnology. Pergamon Press, Oxford, p 183
339. Fowler GL, Higgs Jr RE, Clapp DL, Alford Jr JS, Huber FM (1992) In: Karim MN, Stephanopoulos G (eds) Modeling and control of biotechnical processes. IFAC Symp Ser Pergamon Press, Oxford, p 173
340. Haken H (1988) Condensed Matter 70:121
341. Fuchs A, Haken H (1988) Condensed Matter 71:519
342. Locher G, Sonnleitner B, Fiechter A (1990) Bioproc Eng 5:181
343. Gollmer K, Posten C (1995) In: Munack A, Schügerl K (eds) 6th Internat Conference on Computer Applications in Biotechnology. Pergamon Press, Oxford, pp 41
344. Ignova M, Glassey J, Ward AC, Montague GA, Invine TS (1995) In: ref 292 p 53
345. Wang S, Zou W, Qu H, Yang Y (1995) In: Munack A, Schügerl K (eds) 6th Internat Conference on Computer Applications in Biotechnology. Pergamon, Oxford, p 113
346. Stephanopoulos G, Locher G, Duff M (1995) In: Munack A, Schügerl K (eds) 6th Internat Conference on Computer Applications in Biotechnology. Pergamon, Oxford, p 195
347. Manecke G, Singer S (1959) Macromol Chem 36:119
348. Levin Y, Pecht M, Goldstein L, Katchalski E (1964) Biochem 3:1905
349. Mosbach K, Mosbach R (1966) Acta Chem Scand 20:2807
350. Johnson DE, Ciegler A (1969) Arch Biochem Biophys 130:384
351a. Mosbach K, Larsson PO (1970) Biotechnol Bioeng 12:19
351b. Larsson PO, Ohlson S, Mosbach K (1976) Nature 363:796
352. Toda K, Shoda M (1975) Biotechnol Bioeng 17:481
353. Kiersten M, Bucke C (1977) Biotechnol Bioeng 19:387
354. Tanaka A, Yashukara S, Gelff G, Osumi M, Fukui S (1978) Eur Appl Microbiol Biotechnol 5:17
355. Wada M, Kato J, Chibata I (1979) Eur J Appl Microbiol Biotechnol 8:241
356. Fukui S, Tanaka A (1982) Ann Rev Microbiol 36:145
357. Chibata I (ed) (1978) Immobilized enzyme research and development. Wiley, New York
358. Ghose TK, Fiechter A, Blakebrough N (eds) (1978) Immobilized enzymes I. Adv Biochem Eng, vol 10. Springer-Verlag, Berlin
359. Ghose TK, Fiechter A, Blakebrough N (eds) (1979) Immobilized enzymes II Adv Biochem Eng. Vol 12. Springer-Verlag, Berlin
360. Nabe K, Izuo N, Yamada S, Chibata I (1979) Appl Environ Microb 38:1056
361. Schnarr WG, Szarek WA, Jones IKN (1977) Appl Environ Microbiol 33:732
362. Ghommidh C, Navarro IM, Durand G (1982) Biotechnol Bioeng 24:605
363. Takata I, Yamamoto K, Tose T, Chibata I (1980) Enzyme Microb Technol 2:30
364. Takata I, Yamanoto K, Tose T, Chibata I (1983) Appl Biochem Bioetchnol 8:31

365. Tramper J, van den Tweel WIJ (1981) Abstr Comm Sec Eur Congr Biotechnol Eastbourne, p 161
366. Constantinides A, Bhatia D, Vieth WR (1981) Biotechnol Bioeng 13:899
367. Yamamoto K, Tosa T, Yamashita K, Chibata I (1976) Eur J Appl Microbiol Biotechnol 3:169
368. Yamamoto K, Tosa T, Chibata I (1980) Biotechnol Bioeng 22:2045
369. Murata K, Kato J, Chibata I (1979) Biotechnol Bioeng 21:887
370. Wada M, Uchida T, Kato J, Chibata I (1980) Biotechnol Bioeng 22:1175
371. Yamamoto K, Sato T, Tosa T, Chibata I (1974) Biotechnol Bioeng 16:1601
372. Wada M, Kato J, Chibata I (1981) Eur J Appl Microbiol Biotechnol 11:67
373. Bandyopadhyay KK, Ghose TK (1982) Biotechnol Bioeng 24:805
374. Wada M, Kato J, Chibata I (1980) Eur J Appl Microbiol Biotechnol 10:275
375. Lee TH, Ahn JC, Ryu DDY (1983) Enzyme Microb Technol 5:41
376. Williams D, Münneke DM (1981) Biotechnol Bioeng 23:1813
377. Linko Y, Linko P (1981) Biotechnol Lett 3:21
378. Holeberg IB, Margalith P (1981) Eur J Appl Microbiol Biotechnol 13:133
379. Klein J, Kressdorf B (1983) Biotechnol Lett 5:497
380. Margaritis A, Bajpai, PK, Wallace JB (1981) Biotechn Lett 3:613
381. Slininger PJ, Bothast RJ, Black LT, McGhee JE (1982) Biotechnol Bioeng 24:2241
382. Arcuri EJ, Nichols JR, Brie TS, Santamarina VG, Buckland BC, Drew SW (1983) Biotechnol Bioeng 25:2399
383. Decottignies-Le Marechal P, Calderon-Seguin R, Vandecasteele JP, Azerad R (1979) Eur J Appl Microbiol Biotechnol 7:33
384. Deo YM, Gaucher GM (1984) Biotechnol Bioeng 26:285
385. Morikava Y, Karube I, Suzuki S (1979) Biotechnol Bioeng 21:16
386. Morikava Y, Karube I, Suzuki S (1980) Biotechnol. Bioeng 22:1015
387. Klein J, Hackel U, Wagner F (1979) ACS Symp Ser, vol 106. ACS, Washington DC, p 1001
388. Nilsson I, Ohlson S, Häggstrom L, Molin N, Mosbach K (1980) Eur J Appl Microbiol Biotechnol 10:261
389. Mosbach K, Birnbaum S, Hardy K, Davies J, Bülow L (1983) Nature 302:543
390. Kokubu T, Karube I, Suzuki S (1978) Eur J Appl Microbiol Biotechnol 5:23
391. Kokubu T, Karube I, Suzuki S (1981) Biotechnol Bioeng 23:29
392. Kopp B, Rehm HJ (1983) Eur J Appl Microbiol Biotechnol 18:257
393. Lambe CA, Reading A, Roe S, Rosevear A, Thompson AR (1982) Enzyme Eng 6:137
394. Furusaki S, Seki M (1992) Adv Biochem Eng Biotechnol 46:161
395. Shibatani T (1996) Industrial application of immobilized biocatalysts in Japan In: Wiffels RH, Buitelaar RM, Bucke C, Tramper J (eds) Immobilized cells: Basics and applications. Elsevier, Amsterdam, p 585
396. Kasche V, Lundquist H, Bergman R, Axén R (1971) Biochem Biophys Res Commun 45:615
397. Kasche V (1973) Studia Biophysica 35:45
398. Hooijmans CM, Briasco CA, Hung J, Graats BGM, Barbotin JN, Thomas D, Luyben, KChAM (1990) Appl Microbiol Biotechnol 33:61
399. Huang J, Hooijmans CM, Briasco CA, Graats SGM, Luyben KChAM, Thomas D, Barbotin IN (1990) Appl Microbiol Biotechnol 33:619
400. Furusaki S (1990) Intradiffusion effect on reactivity of immobilized microorganisms. In: Fiechter A et al. (eds) Bioproducts and Bioprocesses. Springer, Heidelberg
401. Doran PM, Bailey JE (1986) Biotechnol Bioeng 28:73
402. Nasri M, Sayadi S, Barbotin JN, Thomas D (1987) J Biotechnol Bioeng 6:147
403. Kumar PKR, Schügerl K (1990) J Biotechnol 14:255
404. Kann JK, Shuler ML (1976/77) An immobilized whole cells hollow fiber reactor for urocanic acid production . AIChE Symp Ser 172: p 31
405. Kann JK, Shuler ML (1978) Biotechnol Bioeng 20:217
406. Pirt SJ, Kurowski W (1970) Gen Microbiol 63:357
407. Portno AD (1967) J Inst Brew 73:43

408. Margaritis A, Wilke CR (1978) Biotechnol Bioeng 20:727
409. Rogers PL, Lee KJ, Tribe DE (1980) Process Biochem 15:7
410. Nishizawa Y, Mitani Y, Tamai M, Nagai S (1983) J Ferment Technol 61:599
411. Chang HN, Furusaki S (1991) Adv Biochem Eng 44:27
412. Bauer S, White MD (1976) Biotechnol Bioeng 18:839
413. Mori H, Kobayashi T, Shimizu S (1979) J Chem Eng 12:313
414. Gleiser IE, Bauer S (1981) Biotechnol Bioeng 23:1015
415. Reiling HE, Laurila H, Fiechter A (1985) J Biotechnol 2:191
416. Cutayar JM, Poillon D (1989) Biotechnol Lett 11:155
417. Knorre WA (1989) BFT-Biotechforum 6:20
418. Lee CW, Gu MB, Chang HN (1989) Enzyme Microb Technol 11:49
419. Lee CW, Chang HN (1990) Biotechnol Bioeng 36:330
420. Riesenberg D, Menzel K, Schulz, V, Schumann K, Veith G, Zuber G, Knorre WA (1990) Appl Microbiol Biotechnol 34:77
421. Knorre WA, Deckwer WD, Korz D, Pohl HD, Riesenberg D, Ross A, Danders E, Schulz V (1990) BioEng 5:28
422. Korz DJ, Rinas U, Hellmuth K, Sanders EA, Deckwer WD (1995) J Biotechnol 39:59
423. Märkl H, Zenneck C, Dubach ACH, Ogbonna JC (1993) Appl Microbiol Biotechnol 39:48
424. Wilmer EN (1935) Tissue culture. Methuen's monographs on biological subjects. Methuen, London
425. Dimopoulosos GT, Pritham GH (1951) J Lab Clin Med 37:162
426. Paul J (1972) In: Rothblat GH, Cristofolo VJ (eds) Growth nutrition, and metabolisms of cells in culture, vol 1. Academic, New York, p 1
427. Weymouth C (1972) In: Rothblat GH, Cristofolo VJ (eds) Growth, nutrition, and metabolism of cells in culture, vol 1. Academic, New York, p 11
428a. Earle WR, Schilling EL, Shannon JE (1951) J Nat Cancer Inst 12:179
428b. Earle WR, Bryant JC, Schilling EL (1953/54) Ann NY Acad Sci 58:1000
429. McCoy TA, Whittle W, Conway E (1962) Proc Soc Exp Biol Med 109:235
430. Molin O, Hedén CG (1969) In: Karger S (ed) Prog Immunbiol Standard 3. Basel p 106
431. van Wezel AL (1967) Nature 216:64
432. Capstick PB, Telling RC, Chapman WG, Stewart DL (1962) Nature 195:1163
433. Hayflick L Moorehead PS, Pomerate CM Hsu TC (1963) Science 140:766
434. Karger S (ed) (1987) Develop Biol Standardization. Advances in Animal cell technology: Cell engineering, evaluation and expolitation. 7th ESACT meeting.
435. Spier RE, Griffiths JB, Stephenne J, Crooy PJ (eds) (1989) Adv in animal cell biology and technology for bioprocesses. Butterworths. 9th ESACT meeting
436. Spier RE, Griffiths JB, Meininger B (eds) (1991) Production of biologicals from animal cells in culture. Butterworths-Heinemann, Oxford. 10th ESACT meeting
437. Spier RE, Griffiths JB, Berthold W (eds) (1994) Animal cell technology. Production of today, prospects for tomorrow. 12th ESACT meeting, Butterlworth-Heinemann, Oxford.
438. Spier RE (1980) Adv in Biochem Eng 14:119
439. Spier RE (1991) In: Ho CS, Wang DIC (eds) Animal cell bioreactors. Butterworth-Heinemann, Stoneham p 3
440. Kelley BD, Chiou TW, Rosenberg M, Wang DIC (1993) In: Stephanopoulos G (ed) Vol 3: Bioprocessing. In: Rehm, HJ, Reed G, Pühler A, Stadler P (eds) Biotechnology, VCH, Weinheim, p 23
441. Fiechter A (ed) (1987) Vertebrate cell culture I. Adv Biochem Eng Biotechnol 34. Springer, Berlin
442. Fiechter A (ed) (1988) Bioprocesses including animal cell culture. Adv Biochem Eng Biotechnol 37. Springer, Berlin
443. Fiechter A (ed) (1980) Plant cell cultures I. Adv Biochem Eng Biotechnol 16. Springer, Berlin
444. Fiechter A (ed) (1980) Plant cell cultures II. Adv Biochem Eng Bioetchnol 18. Springer, Berlin
445. Ratafia M (1987) Bio/Technol 5:1154

446. Croughan MS, Wang DIC (1989) Biotechnol Bioeng 33:731
447. Wudtke M, Schügerl K (1987) Investigations of the influence of physical environment on the cultivation of animal cells. In: Spier RE, Griffits JB (eds) Modern approaches to animal cell technology. Butterworth, London, p. 297
448. Michaels JD, Mallik AK, Papoutsakis ET (1996) Biotechnol Bioeng 51:399
449. MacLoughlin PF, Malone DM, Murtagh JT, Kieran PM (1998) Biotechnol Bioeng 58:595
450. Nilsson K, Mosbach K (1979) FEBS Lett 118:145
451. Nilsson K, Scheirer W, Merten OW, Ostberg L, Liehl E, Katinger HWD, Mosbach K (1983) Nature 302:629
452. Brodelius P, Nilsson K (1980) FEBS Lett 122:312
453. Brodelius P, Mosbach K (1982) Adv Appl Microbiol 28:1
454. Bierbaum S, Larsson PO, Mosbach K (1986) Immobilized biocatalysts. The choice between enzymes and cells. In: Webb C, Black GM, Atkinson B (eds) Process engineering aspects of immobilized cell systems. The Inst Chem Engrs, Warwickshire
455. Heath C, Belfort G (1987) Adv Biochem Eng Biotechnol 34:1
456. Lim F, Moss RD (1981) J Pharm Sci 70:351
457. Jarvis AP, Grdina TA (1983) Biotechniques 1:22
458. Knazek RA, Gullino PM, Kohler PO, Dedrick RL (1972) Science 178:65
459. Knazek RA, Kohler PO, Gullino PM (1974) Exp Cell Res 84:251
460. Merten OW et al. (eds) (1998) Animal Cell Technology: New Developments, New Applications. Kluwer Academic Press. 15th ESACT meeting 1997, Tours, France

Received December 1998

A View of the History of Biochemical Engineering

Raphael Katzen[1], George T. Tsao[2]

[1] 9220 Bonita Beach Road, Suite 200 Bonita Springs, Florida 34135, USA
[2] School of Chemical Engineering, Purdue University, West Lafayette, Indiana 47907, USA

The authors present a view of biochemical engineering by describing their personal interests and experience over the years involving mostly conversion of lignocellulosics into fuels and chemicals and the associated engineering subjects.

Keywords. Biomass conversion, Biochemical engineering, Fuels, Chemicals, History.

1 Introduction 77

2 Early Development of Biochemical Engineering 78

3 Early Development on Conversion of Lignocellulosics 80

3.1 Concentrated Acids and Solvents 80
3.2 Dilute Acids 81
3.3 Enzymes 82

4 Renewed and Expanded Efforts on Biomass Conversion 82

5 Further Advances in Biochemical Engineering 86

6 Further Advances in Biomass Conversion 87

7 Concluding Remarks 89

References 89

1 Introduction

Biochemical engineering has grown into a very broad subject field. The scope of this article is limited mostly to technology for conversion of lignocellulosic biomass into fuels and chemicals, and the associated biochemical engineering topics. The content reflects the interests or personal experience of the authors. It offers a limited view of the history of biochemical engineering. History, as always, has to be told from many different viewpoints, to achieve an objective and complete exposition.

The phrase, "biochemical engineering", first appeared in the late 1940s and early 1950s. That was the time shortly after aerobic submerged culture was

Advances in Biochemical Engineering/
Biotechnology, Vol. 70
Managing Editor: Th. Scheper

launched as a way of increasing the production capacity of penicillin, used to cure battle wounds of World War II (Shuler and Kargi 1992). Fungal mycelia grow naturally on the surface of moist substrates. When mycelia are submerged in liquid nutrients, an adequate supply of oxygen, often in the form of finely dispersed air bubbles to support aerobic biological activities, has become an important requirement. Gas-liquid interfacial mass transfer of oxygen in reaction vessels has since become a challenge to those trained in chemical engineering. An article by Hixon and Gaden (1950) on oxygen transfer in bioprocesses initiated a wave of activities that has often been credited as the first recognition of "biochemical engineering" as an engineering subject requiring systematic studies to understand its governing principles and for acquisition of skills for good design and performance.

Commercial scale biological processing of biomass materials is an activity as old as human civilization. In more recent years, utilization of lignocellulosic biomass has become an actively pursued subject, because of the concerns of future exhaustion of non-renewable fossil fuels. One of the authors of this article, Raphael Katzen, had his first personal experience in wood hydrolysis 60 years ago in the late 1930s (Katzen and Othmer 1942). Since the energy crisis caused by the oil embargo by the OPEC countries in 1974, there have been expanded efforts in research and development aimed at improved conversion efficiency of lignocellulosics as an alternative material resource. The future payoff from successful utilization of lignocellulosics will be enormous. In fact, the future of human society may depend on it, which might not be so obvious to many business leaders and policy makers today, but it will become increasingly clear in the years to come. Most of the lignocellulosic materials today are considered "wastes" or, at best, low value materials. In order to achieve profitable industrial scale conversion of lignocellulosics into chemicals, materials and fuels, concerted efforts of scientists and engineers from many disciplines, including biochemical engineering, are needed.

2 Early Development of Biochemical Engineering

Following the commercialization of penicillin, a large number of antibiotics were also discovered from extensive screening programs by many pharmaceutical companies worldwide in the 1950s and 1960s. There was a strong demand for better designs of aeration systems and deeper understanding of the process of oxygen transfer in biological systems so that many new drugs could be manufactured efficiently. Oxygen transfer became in those years a popular subject for biochemical engineers to engage in. One of the authors of this article, George T. Tsao, spent his early career pursuing this topic, starting with his own graduate thesis. In those years, one of the most important references in oxygen transfer is the comprehensive review by Professor Robert Finn (1954). Two other frequently cited articles on oxygen transfer in bioreaction systems include the one by Hixon and Gaden mentioned above and another one by Bartholomew, Karow, Sfat and Wilhelm. Both, in fact, appeared in the same Fermentation Symposium in 1950.

In studying oxygen transfer, the first problem at the time, was how one could measure its rate. This problem led to another most frequently cited reference by Cooper, Fernstrom, and Miller (1944), where the rate of oxidation of sulfite ions to sulfate ions was described as a method to reflect the rate of oxygen transfer into gas-liquid contactors. While aerobic processes was flourishing in the biochemical industry, aerobic and anaerobic wastewater treatment were also becoming increasingly an important and widely used processes in the sanitation industry. The fundamentals are the same, whether it is a bioreaction medium or a liquid waste, both requiring adequate dissolved oxygen in some stages to support microbiological activities. There was considerable interaction among biochemical engineers and sanitary engineers. For instance, among the early references important to oxygen transfer is the publication edited by McCabe and Eckenfelder (1955) on "Biological Treatment of Sewage and Industrial Wastes."

While the field of biochemical engineering was growing in its infancy, the field of chemical engineering was maturing in the 1960s. The famous "Bird" book on Transport Phenomena (Bird, Stewart, and Lightfoot 1960) started to place a solid scientific foundation underneath many of the chemical engineering processes and operations. Meanwhile, "Chemical Reaction Engineering" had evolved from its chemical kinetics origin into a full and important branch of chemical engineering. The book on this subject by Levenspiel (1962) helped educate several generations of chemical engineers. Ever since, design and performance of chemical reactors as well as bioreactors have had systematic engineering guidelines and principles.

Often, biotechnologists get excited when they find certain super microorganisms capable of synthesis and accumulation of a valuable metabolite. Soon, they realize that the product cannot be marketed and it has to be purified to meet necessary specifications. Bioseparation, a phrase coined much later in the 1980s, also started to become an important branch of biochemical engineering. The early work involved mostly adopting separation techniques such as solvent extraction and crystallization, well developed in chemical process industries, to purify biochemical products such as antibiotics, organic acids, vitamins, and others. It was later in the 1980s, when chromatography, membrane separation, electrophoresis, super centrifugation, and so on, were needed for purifying many protein and other sensitive biological products, that bioseparation started to become an important engineering factor.

In 1959, a new journal entitled "Biotechnology and Bioengineering" was first published by John Wiley, with Professor Elmer L. Gaden as its founding managing editor. This journal has since become one of the most important publications in biochemical engineering. In those days, the word biotechnology meant simply the technology based on activities of biological and biochemical materials. This word still means the same to many people today. However, there are now also many who interpret this word solely as activities related to genetic modification of living systems.

While aerobic processes had a close association with the first coinage of the phrase "biochemical engineering"; anaerobic bioprocesses actually have a long and important history. Wine, rum and whisky making involving ethanol fermentation have always been the most important bioprocess, where, as every one knows, a large oxygen supply is not desired. Methane generation by degradation of biomass materials, occurring naturally or man-made, is also one of the most important anaerobic processes. Without the anaerobic digestion in the bodies of rumen animals, the meat and agricultural industries would have been very different from what we have today. Lactic acid bioprocesses for either the product or for making silage are again carried out by anaerobic microbes. Many other anaerobic processes are also involved in the preparation of a variety of indigenous foods in different countries.

Besides lactic acid, one of the processes that did become fully industrialized was the bacterial anaerobic bioproduction of solvents (Underkofler and Hickey 1954). For some years, the anaerobic process was the main source of the industrial solvents including acetone and butanol, until the rise of the new petrochemical industry. This strictly anaerobic process now no longer exists except in a few laboratory studies.

3
Early Development on Conversion of Lignocellulosics

The annual production of biomass is 60 billion tons worldwide. Waste biomass in the United States is one billion tons per year. If it can be converted into chemicals and materials, human population will be able to enjoy material abundance for ages to come. Biomass materials are renewable. Their utilization creates no net gains of greenhouse gases in the atmosphere. The processing methods utilize mild reaction conditions, creating relatively few pollutants. Among biomass, lignocellulosic biomass is currently of relatively little use. For instance, together with annual harvest of about 360 million tons of farm crops such as corn and wheat in the United States, there are co-produced about 400 million tons of lignocellulosics such as cornstalks and wheat straws, often referred to as crop residues. Most of these residues are left in the field for natural degradation, or collected and burned. Lignocellulosics typically contain 70% or more by weight of polysaccharides, including cellulose and hemicellulose. Once they are converted into monosaccharides such as glucose, xylose and others, bioprocessing methods can be applied to convert them into a large number of chemicals and fuels. Hydrolysis of cellulose to produce glucose, however, has not been an easy task. Numerous attempts have been made but economic success, even today, is still limited. Processes for cellulose hydrolysis can be roughly divided into three categories: concentrated acids and "solvents," dilute acids, and enzymes.

3.1
Concentrated Acids and Solvents

The earliest approach to conversion of the carbohydrate fraction to sugars stems from the more than 100 year old Klason lignin determination, in which hemicellulose and cellulose are gelatinized in 72% sulfuric acid, and after dilution with water, hydrolyzed to yield mixed five-carbon and six-carbon sugars. The residue from this solubilization and hydrolysis of the carbohydrate fractions leaves a residue identified as lignin (TAPPI 1988). Although this residue could contain other materials such as wood oils and ash, modern chromatographic analysis permits identification of the true lignin content of this residual fraction. Attempts were made in recent years to use this analytical procedure as a pretreatment to gelatinize cellulose and hemicellulose, and then hydrolyze it to obtain a high yield of sugars.

Another concentrated acid process was developed in Germany prior to World War II (Bergius 1933). This process utilized 40% hydrochloric acid for the solubilization stage followed by dilution to complete the hydrolysis. Here, recovery of the large amounts of costly hydrochloric acid is essential. However, during World War II the primary stage of this technology was utilized in Germany to convert wood waste to sugars, followed by neutralization of the acid with sodium hydroxide to yield a mixture of wood sugars and sodium chloride, suitable for use as cattle feed, as a partial replacement or substitute for limited and costly grain feeds (Locke 1945).

3.2 Dilute Acids

Prior to World War II, technology was also developed in Germany (Scholler 1935), utilizing a dilute sulfuric acid percolation process to hydrolyze and extract pentoses and hexoses from wood waste. Several installations were built in Germany prior to and during the war. It is estimated that about 50 Scholler type installations were built in the former Soviet Union during and after World War II. Some of the wood hydrolysates were processed by yeast to produce cell mass in what is called the Waldhof system in Germany. Draft tubes were installed to induce air dispersion into the reaction mixture: to supply dissolved oxygen to growing cells. Draft tube aerators similar in design to the original Waldhof system are now still widely in use in bioreactors and also in aerobic wastewater treatment facilities. When Leningrad of the former Soviet Union (now St. Petersburg of Russia) was under siege for two years by the advancing German army during World War II, hydrolysis of lignocellulosics was used as a source of some digestible carbohydrates. There is a Hydrolysis Institute in that city. Its war-time director earned two Lenin Medals, the highest honor in those days.

One plant was also built at Domat-Ems, Switzerland by Holzverzuckerrungs A.G., employing the dilute sulfuric acid method. A substantial facility was designed, built and operated under direction of Raphael Katzen for the Defense Plant Corporation of the U.S. Government during World War II at Springfield, Oregon for processing 300 tons per day of sawdust from nearby sawmills, yielding 15.000 gallons per day of ethanol, utilizing *Saccharomyces* for the process (Harris 1946). This yield of 50 gallons per ton of wood was approximately 50% of the theoretical yield. The indicated loss of sugars and production of furfural from the pentoses, as well as possible reaction with lignin, resulting in formation of tarry residues which, when mingled with calcium sulfate derived from neutralization of the sulfuric acid, resulted in major scaling and blockage problems. After full-scale production proved capacity and nameplate production, the plant was shot down as being uneconomic after the war in competition with rapidly developing low-cost synthetic ethanol.

Research on improvement of the dilute sulfuric acid process continued at the Tennessee Valley Authority after World War II (Gilbert 1952). Dilute sulfuric acid hydrolysis of wood and other lignocellulosic materials was also investigated at the Forest Products Laboratory of the U.S. Department of Agriculture, in Madison, Wisconsin. A number of publications (Saeman 1945) from these efforts become important references on the subject in later years. Despite all efforts, ethanol yields from wood with dilute sulfuric acid technology did not exceed 60% of theoretical.

3.3 Enzymes

Again, it was the war-time efforts by researchers at the U.S. Army Quartermaster Research Center in Natick, Massachusetts, that led to the discovery of cellulose hydrolyzing enzymes, commonly known as cellulases. It was told that army uniforms made of cotton were biodegraded quickly in tropical places during World War II. Under the leadership of Elwyn T. Reese and Mary Mandels (1975), cellulases were identified as the cause of degradation of cellulose in cotton fabrics. The culture that was isolated as one of the potent cellulases producers was *Trichoderma viride* which was later re-named *Trichoderma reesei* in honor of Dr. Reese. The Natick center continued to provide leadership in cellulase research for many years.

4 Renewed and Expanded Efforts on Biomass Conversion

After World War II, there was a long period of prosperity of about two decades. Consumption of petroleum products increased quickly. There were relatively little commercial and research interests in alternatives to petroleum. The oil embargo in 1973–74 served as a wake up call, which renewed strong interest in utilization of alternative resources. Renewable biomass and coal were looked upon for possible replacement of fuels and chemicals from petroleum. At the time, George Tsao was on assignment at the U.S. National Science Foundation, on leave from Iowa State University, managing several funding programs as a part of the RANN (Research Applied to National Needs) initiatives. NSF supported work on organizing conferences and workshops to identify research needs. Recognizing the need for biomass research, George Tsao invited Professor Charles R. Wilke of the University of California at Berkeley to conduct a conference on the use of cellulose as a potential alternative resource of fuels and chemicals. It took place in 1974 and the proceedings of that conference were later published as a special volume of Biotechnology and Bioengineering (Wilke 1975). That conference served an important function in stimulating renewed interest in the conversion of biomass into fuels and chemicals. In 1978, the first of a series of conferences on Biotechnology for Fuels and Chemicals (first it was named Biotechnology in Energy Production and Conservation, but later revised to cover other chemicals also) was organized by researchers of the Oak Ridge National Laboratory, Oak Ridge, Tennessee, under the leadership of Dr. Charles D. Scott. Later, researchers of the National Renewable Energy Laboratory, Golden, Colorado also joined the effort and this series of conferences has been held annually ever since. In May 1999, two hundred scientists and engineers from many countries attended the 21st Conference held in Fort Collins, Colorado. The conference proceedings were first published as special issues of B&B and later by the journal, Applied Biochemistry and Biotechnology. This series of publications has turned out to be probably the most important information and reference source in this field.

Attempts of applying the concept of first gelatinizing cellulose and then hydrolyzing it to produce degradable sugar and then ethanol have been carried out over the years. In the late 1970s, researchers at Purdue University further investigated that concept by using concentrated sulfuric acid, concentrated hydrochloric acid, together with several other "cellulose solvents" such as Cadoxin (Tsao 1978), resulted in the issue of several patents. The concentrated sulfuric acid method was investigated again at the Tennessee Valley Authority (Farina 1991) and the U.S. National Renewable Energy Laboratory (Wyman 1991). Recent work by ARKENOL in California, USA (Cuzen 1997) and APACE in New South Wales, Australia, have hinged on developing novel, economical methods for separation of sugars from acid, and recovery of substantial amounts of diluted sulfuric acid, evaporated to the required concentration for recycle to the gelatinization stage of the process. Both membrane and ion-exclusion technologies have been tested and developed, toward eventual demonstration and commercialization of the separation techniques. During the process of interacting concentrated sulfuric acid and cellulose, there are likely chemical reactions taking place between the acid and the substrates, which could influence the yield as well as the degradability of the monosaccharides so obtained as well as the acid recovery. This possibility should be carefully investigated before the process can be commercialized. The use of concentrated hydrochloric acid first applied in the World War II era for conversion of cellulose to sugars was again investigated by Battelle-Geneva on a pilot plant basis, particularly of separation of the hydrochloric acid and sugars, as well as re-concentration of the hydrochloric acid for recycle.

The early work on dilute acid hydrolysis was also revisited at Purdue University (Ladisch 1979), in New Zealand (Whitworth 1980) and recently at NREL (Wyman 1992). The use of dilute acids under mild reaction conditions were looked upon by those at Purdue to serve two functions: removal of hemicellulose and pretreatment for cellulose hydrolysis. Under very mild reaction conditions, dilute sulfuric acid removed hemicellulose to form a hydrolysate containing mostly xylose, arabinose and other hemicellulose sugars, without attacking the cellulose fraction in the substrate. After removal of the hydrolysate, cellulose left in the solid residues was then subjected to either dilute acid hydrolysis at a higher temperature or treated with a concentrated sulfuric acid for gelatinization and then hydrolysis. This 2-stage acid Purdue Process generated two sugars streams (Tsao, Ladisch, Voloch and Bienkowski 1982).

Because of the mild reaction conditions, the hemicellulose hydrolysate containing xylose was not contaminated by furfural and other degradation products that would inhibit microbial activities in subsequent bioprocesses. The glucose stream from cellulose was not contaminated by pentoses making its utilization straightforward.

There were also renewed and widespread interests in enzymatic hydrolysis of cellulose using cellulases. Researchers at the Rutgers University conducted a successful culture mutation program (Montenecourt and Eveleigh 1977 1978). A super-productive *Trichoderma reseei* RUT C-30 strain was resulted from it as a mutant of the then best enzyme producer, QM9414, from Natick. This C-30 culture has continued to be one of the best cellulase producers even today. Meanwhile, modern enzymology techniques were extensively applied to investigate the properties and the reaction kinetics of cellulases (Gong 1979) leading to much better understanding of how these enzymes work symbiotically in converting cellulose into glucose.

Developments initiated by the Bio-research Corporation of Japan, a partnership of Gulf Oil and Nippon Mining, resulted in two basic patents, one pertaining to production of the cellulase enzyme (Huff 1976), while the other initiated the principle of simultaneous saccharification and degradation (SSF) (Gauss 1976). This milestone invention overcame the very long saccharification period due to glucose feed back inhibition of the cellulase activities. Major improvement developed by Gulf Oil Chemical Research Group (Emert 1980) at the Shawnee, Kansas Research Laboratory, and later by the same group after transfer to the University of Arkansas, included a continuous 48 hour process for production of cellulase enzymes, as well as a method of recycling active enzyme from the resulting fermented beer by adsorption on fresh feedstock (Emert 1980).

One of the recognized research needs in biomass conversion was the use of 5-carbon sugars derived from hemicellulose for improvement of overall process efficiency. Once glucose is obtained from cellulose hydrolysis, there is no fundamental problem of making good use of it. Xylose derived from hemicellulose is a different matter. Most good glucose-degrading yeast cells cannot metabolize xylose. An important breakthrough made by Dr. C.S. Gong of Purdue University was the conversion of xylose into its isomer, xylulose, using a commercial enzyme, glucose isomerase. Xylulose can then be readily fermented by *Saccharomyces* yeast to ethanol (Gong 1981 and 1984).

The mid-1970s was also the time when gene splicing was made straightforward because of the application of restriction enzymes and other advances in molecular biology. The phrase, Genetic Engineering, was coined at that time. After C.S. Gong's discovery, Nancy Ho was appointed to head a Molecular Genetics Group in the Laboratory of Renewable Resources Engineering (LORRE) of Purdue, with the main objective of splicing isomerase gene into *Saccharomyces* yeast cells. The work did not succeed for almost ten years. Later, the effort was re-directed to replace isomerase gene with two genes: one for a reductase and another one for a dehydrogenase for converting xylose first to xylitol and then xylulose. In addition, a gene coded for the xylulose kinase was also transferred into *Saccharomyces* to build a new metabolic pathway to convert xylose into ethanol via the pentose phosphate cycle. This effort, as well known today, resulted in success (Ho 1997). This long effort started in the late 1970s and was metabolic engineering in nature long before the phrase, "metabolic engineering," was coined in the early 1990s.

There were several parallel attempts worldwide in searching for microbial cultures capable of degrading xylose to ethanol. The search includes both the use of natural occurring cultures and also genetically engineered ones. Most notable was the work at the University of Florida led by Professor L. O. Ingram, which was made famous by the issuance of the US Patent 5,000,000 (1991). The Ingram group created recombinant *Escherichia coli* and an improved recombinant *Klebsiella oxytoca*, achieving 98% of theoretical conversion of five carbon sugars to ethanol. Work being done at the University of Toronto (Lawford 1998) utilizing *Xymomonas mobilis* also appears promising, along with work by others (Wilkinson 1996). The genetically engineered *Saccharomyces* was successfully tested at the NREL Process Development Unit at Golden, Colorado. Since then, new *Saccharomyces* cultures have been created by Dr. Ho's group, with the above mentioned foreign genes fused into the chromosomes, assuring improved genetic stability.

A challenging problem associated with xylose conversion to ethanol by either natural or genetically modified microbes is really in the rate of fermentation. Often the biomass hydrolysates contain both glucose and xylose and possibly other monomeric components. In the process, glucose will quickly be exhausted but the process may require another day or two to complete xylose conversion. In other words, the xylose-degrading capability will increase ethanol yield from lignocellulosics with a larger sugar basis, but the slow xylose turnover rate may actually decrease the productivity of the fermentation vessel, which is usually expressed in the amount of ethanol produced per unit time per unit volume of the vessel. Future work on xylose conversion to ethanol should include this issue under careful consideration to bring true overall process improvements.

When the efforts on gene splicing for xylose conversion to ethanol were taking place, parallel efforts were also made worldwide for conversion of xylose and other sugars into chemicals other than ethanol. Over the years extensive eforts have been made on production of butanediol, furfural, xylitol, lactic acid, SCP from pentoses. Meanwhile, researchers worldwide also branched out from its early concentration on renewable fuels of ethanol and methane to other chemicals including acetic acid, lactic acid, glycerol, fumaric acid, citric acid, malic acid, succinic acid, aspartic acid, bacterial polysaccharides, acetone, butanol, butyric acid, methyl ethyl ketone, just to mention a few in a partial list. If lignocellulosics-based chemical industries are ever to compete effectively and eventually replace crude oil-based chemical industries, integrated synthetic networks are needed to rival the complexity and sophistication of the current petrochemical synthetic networks. Manufacture of ethanol alone as a single product of a processing plant is unlikely to be economically effective, without crediting the fuel energy value of the residues. By-products and co-products of ethanol should be expected in future large processing and chemical manufacturing enterprises based on lignocellulosics feedstocks. This view seems to be also held by many researchers in this field. A quick search of the literature in the last 20 years, or a look at some 200 articles presented at the recent 21st Conference on Biotechnology for Fuels and Chemicals will give an unmistakable impression that scientists and engineers worldwide have been and still are pursuing actively many other products from biomass in addition to ethanol.

5
Further Advances in Biochemical Engineering

The required engineering expertise in manufacture of ethanol and bulk chemicals is somewhat different from that in producing high priced health products. For low value products, the required efficiency is important in determining the overall process economics. For pharmaceuticals, assurance of the product purity and safety is of top priority. The price of a new drug can always be properly adjusted to give manufacturers the desired profit. For products like ethanol, process efficiency requirements stimulated a great deal of engineering research and development work over the years. In the late 1960s and early 1970s, there was a worldwide perception of the possible shortage of food to feed the world population. The oil companies placed strong emphasis on conversion of hydrocarbons into single cell proteins. The demand of dissolved oxygen to support cell growth with hydrocarbons as the carbon sources is even stronger than that when carbohydrates are the substrate. The problem of oxygen transfer and cell growth also led to a number of new and low cost designs of fermentation vessels. Airlift reactors with either internal or external circulation loops were tested on a large scale (Schuler and Kargi 1992). At that time, dissolved oxygen analyzers were built and first marketed in the late 1960s. With this instrument, the control of adequate supply of dissolved oxygen became much easier. The above mentioned sulfite oxidation method often led to wrong conclusions when the reaction conditions are not well understood and properly controlled (Danckwerts 1970). With a dissolved oxygen analyzer, not only the DO concentration but also the rates of oxygen input into the reaction mixture as well as uptake by cells can be readily determined with dynamic measuring procedures (Mukerjee 1972, Tsao 1968). The use of oxygen analyzers should be considered a milestone advance in the history of biochemical engineering. The interests in conversion of hydrocarbons into SCP quickly disappeared after the oil embargo in 1974. Soon afterwards, with the advances made in DNA recombinant technology and the energy crisis in the mid-1970s, the emphasis of biochemical engineering changed to work on growth of animal cell cultures and purification of proteins and other modern health products as well as the conversion of biomass into chemical and fuel products.

In addition to the above mentioned SSF (simultaneous saccharification and degradation) process, new methods of simultaneous degradation and product recovery (SFPR) processes were focal points of extensive investigation (Cen 1993). Biological agents almost always suffer from product feedback inhibition. The SSF concept avoids the feedback inhibition of cellulase activities. SFPR or SSFPR will help to reduce the feedback inhibition, for instance, of ethanol on yeast cells. For ethanol, a number of techniques were proposed and tested, including, for instance, membrane separation of the ethanol product (Matsumuro 1986), evaporation of ethanol from processing beer by gas stripping (Dale 1985, Zhang 1992), pervaporation (Mori 1990), solvent extraction (Bruce 1992), and others. Meanwhile, there has been growing interest, as mentioned above, in producing other chemicals from degradable sugars. For acidic products such as lactic acid intended for preparation of polylactide biodegradable plastics, there was work on simultaneous removal of the acid by adsorption on resin columns coupled with the bioreactors (Zheng 1996). Among the many bioprocesses of commercial interest, past or current, the anaerobic acetone-butanol-ethanol process is the most sensitive to product feed-back inhibition. At only about 8 g/l of butanol, the reaction will be stopped completely. This process has a acid-formation phase followed by a solvent production phase. Acetic acid and butyric acid formed in the first phase are later converted into the solvents. Attempts were made to adsorb the acids with selected resins, resulted in two possibilities. One can, in one case, continue to produce the solvents, or one can change the process to producing organic acids instead (Yang 1994a, b 1995).

In order to improve reactor efficiency, multiple stage, fed batch reaction systems with internal cell recycle has been developed for ethanol production in the industry. For reducing energy cost spent in dehydration of ethanol to produce fuel grade products, the technique of adsorption by corn grits was invented (Ladisch 1984). This technique is currently used in industrial production of about 60% of the fuel ethanol in the United States. For increasing reactor productivity, cell recycle to build up to high cell population density has been investigated. In short, many advances in improved engineering of processing techniques and reactor designs have been resulted from a large amount of research and development efforts in the last decade. This type of true engineering work will continue for years to come to support large scale conversion of lignocellulosics into low value chemicals and fuels where high processing efficiency is of critical importance.

Even though the concept of metabolic engineering was not new, the work in this area took off and expanded greatly in the 1990s. With the tools of DNA recombinant technology and protein engineering well in hand, metabolic engineering promises to create many new opportunities in biomass conversion. The new techniques of DNA chips, microarray, and combinatorial techniques will bring biochemical engineering to yet another high level of sophistication, greatly benefiting the biomass technology.

6 Further Advances in Biomass Conversion

Gelatinization by concentrated sulfuric acid was the first technique of pretreatment of wood before high yields of glucose could be obtained in the subsequent hydrolysis (Cuzens 1997). Since the mid-1970s, several other pretreatment techniques have been investigated with varying degrees of success. The results from enzymatic hydrolysis reported by the Natick group were derived from lignocellulosic substrates after ball milling as a pretreatment. Extensive ball milling apparently can decrystallize cellulose and making it more accessible to enzymatic hydrolysis. However, the large energy expense in balling milling makes the method impractical. Steam explosion is another pretreatment method being extensively investigated (Schell 1991). The high steam temperature associated with the high pressure degrades large fractions of pentoses from hemicellulose making the pretreatment less desirable. Another explosion method was then invented involving the use of compressed ammonia replacing high pressure, high temperature steam. The low temperature ammonia achieves explosion without the side effect of the high temperature in steam explosion (Holtzapple 1991). In this process, a high level of ammonia recovery is needed to reduce chemical cost. Yet, another method involving carbon dioxide explosion at low temperature has been investigated (Zheng 1995, 1996). An interesting side benefit observed in carbon dioxide explosion is the sharp decrease in cellulose crystallinity by the pretreatment, in addition to exploding the substrate into fine powders. It has been speculated that carbon dioxide being hydrophobic, unlike steam, is capable of entering the crystal lattice of cellulose and causing disruption.

With pretreatment techniques and bioreactor designs well advanced, product recovery, as well as production of many other chemicals besides ethanol being extensively investigated, the entire field of biomass conversion is on the verge of becoming a real competitor in the chemical manufacturing industry. In more recent years, increasing numbers of chemicals that may have been products of petrochemical processing, are being produced by biomass conversion. One well known example is lactic acid (Du 1998). Fumaric acid is another organic acid for which biomass conversion may become the method of industrial production, replacing petrochemical synthesis (Cao 1996). There is now a very efficient method of production of glycerol by degradation of sugars, which may soon become highly competitive against chemical synthesis of this compound (Gong 1999). Many researchers worldwide are searching for a new bioprocess to produce 1,3-propanediol, a monomer needed in the synthesis of new classes of polyesters. The list of such chemicals is growing and the rate may accelerate before long.

Meanwhile, extensive discussion among experts in the field has identified the high cost of cellulases in the current market as an important roadblock to large scale conversion of lignocellulosics. Cellulase products in the current market are mainly designed for application in textile treatment instead of biomass conversion to chemicals. Textile industry, by nature, can tolerate a higher expense of enzymes than the future ethanol industry.

To search for methods of decreasing the production cost of cellulases, recent work has discovered that solid phase processing (SPF) might be useful (Tsao 1999). If oxygen transfer is the most important engineering problem when the liquid submerged fermentation was first introduced, heat dissipation and also oxygen transfer are the main engineering problems in large scale SPF. The work on improving heat dissipation from and oxygen transfer into porous packed beds of moist solid substrates is just beginning, and much additional systematic work is needed to resolve the difficulties. If SPF can be made to perform well, a new era of bioprocess technology might be born much like the period, from

1960s and on, being the new era of submerged processing mode. SPF is definitely not new because making rice wine, sorghum liquor, and soy sauce in the Far East since ancient times has been done by solid phase processing mode. The old technique of SPF needs new improvements to meet the requirements of the modern society. Indeed, there is a great potential in fulfilling them.

7 Concluding Remarks

As stated above, this article is limited to biochemical engineering and biotechnology applied to the conversion of lignocellulosics into fuels and chemicals. Large and important portions of modern biochemical engineering dealing with health products, bioremediation of toxic wastes, and others are left out for other experts to describe.

References

Barthelomew WH, Karow EO, Sfat MR, Wilhelm RH (1950) Ind Eng Chem 42:1801
Bartisch CM (1979) Ethanol by Homologation of Methanol, US Patent 4, 171, 461
Bergius F (1933) Trans Inst Chem Eng (London) 11:162
Bird RE, Stewart WE, Lightfoot EN (1960) Transport Phenomena. Wiley, NY
Bruce LJ, Daugulis AJ (1992) Extractive Fermentation by *Zymomonas mobilis* and the Use of Solvent Mixtures, Biotechnol Lett 14:71
Cantarella M et al. (1991) Enzymatic Hydrolysis of Biomass, International Symposium for Alcohol Fuels Proceedings 1:149–154
Chang HN, Furusaki S (1991) Membrane Reactors. Adv Biochem Eng/Biotechnol 44:29
Cen P et al. (1993) Recent Advances in the Simultaneous Bioreactions and Product Separation Processes, Separation Technol 3:1–18
Cooper CM, Fernstrom GA, Miller SA (1944) Ind Eng Chem 36:504
Cuzens J et al. (1997) Innovative Method for Separating Cellulose Hydrolysis Products, present at the 19th Symposium on Biotechnology for Fuels and Chemicals, Colorado Springs, USA
Dale MC, Okos MR, Wankat PC (1985) An Immobilized Cell Reactor with Simultaneous Product Separation, Biotechnol Bioeng 27:932
Danckwerts PV (1970) Gas-Liquid Reactions, McGraw-Hill, NY
Delgenes JP et al. (1991) C5 Sugars to Ethanol by *Pichia stipitis*, 9th International Symposium for Alcohol Fuels Proceedings, Vol. 1:190–194
Du J et al. (1998) Production of L-Lactic Acid by *Rhizopus oryzae* in a Bubble Column Fermenter, Appl Biochem Biotechnol 70–72, 323–329
Emert GH (1980) US Patent 4,220,721, Method for Enzyme Reutilization
Emert GH et al. (1980) Economic Update of the Gulf Cellulose Alcohol Process, Chem Eng Progress Sept 47–52
Farina GE et al. (1991) Ethanol from Refuse-Derived Fuel, 9th International Symposium for Alcohol Fuels Vol. 1:277–281
Finn RK (1954) Bacterial Review 18:254
Gauss WF et al. (1976) US Patent 3,990,9444, Manufacture of Alcohol from Cellulosic Materials Using Plural Ferments
Gilbert N et al. (1952) Hydrolysis of Wood, Ind Eng Chem 44:1712–1720
Gong CS et al. (1999) Co-Production of Ethanol and Glycerol, presented at the 21st Symposium on Biotechnology for Fuels and Chmicals, Fort Collins, CO
Gong CS et al. (1981) Production of Ethanol from D-Xylose by Using D-Xylose Isomerase and Yeasts, Appl Env Microbiol 41:430–436

Gong CS, Ladisch MR, Tsao GT (1979) Biosynthesis, Purification and Mode of Action of Cellulases of Trichoderma reesei, Adv Chem Ser No. 181:261–287, American Chemical Society Washington, DC

Harris EE et al. (1946) Madison Wood Sugar Process, Ind Eng Chem 38:896–904

Hixon AW, Gaden EL Jr (1950) Ind Eng Chem 42:1792

Ho NWY et al. (1997) Development of a Recombinant Xylose-Fermenting Enhanced *Saccharomyces*, presented at the 19th Symposium on Biotechnology for Fuels and Chemicals, Colorado Springs

Holtzapple MT et al. (1991) The Ammonia Freeze Explosion (AFEX) Process, Appl Biochem Biotechnol 28–29:59–74

Huff GF et al. (1976) US Patent 3,990,945, Enzymatic Hydrolysis of Cellulose

Ingram LO et al. (1991) US Patent, 5,000,000 Ethanol Production by *Escherichia coli* Strains Co-Expressing *Zymomonas* PDC and ADH Genes

Katzen R (1990) Ethanol from Lignocellulose Agro-Industrial Revolution Conference, Washington, DC

Katzen R (1991) Advancing Technology for Ethanol, 9th International Symposium for Alcohol Fuels Vol. 1:131–136

Katzen R, Othmer DF (1942) Wood Hydrolysis – A Continuous Process, Ind Eng Chem 34:314

Ladisch MR (1979) Fermentable Sugars from Cellulose Residue, Process Biochem 21–24

Ladisch MR et al. (1984) Cornmeal Adsorber for Dehydrating Ethanol Vapors, Ind Eng Chem Process Res Dev 23:437–443

Lawford HG et al. (1998) Continuous Culture Studies of Xylose-Fermenting Zymomonas mobilis Appl Biochem Biotechnol 70–72:353–367

Lievenspiel O (1962) Chemical Reaction Engineering, Wiley, New York

Locke EC et al. (1945) Production of Wood Sugar in Germany, FIAT Final Report, Office of Military Government for Germany, Washington, DC

Mandels M (1974) Enzymatic hydrolysis of waste cellulose. Biotechnol Bioeng Proceedings 16:1471–93

Mandels M (1975) Microbial Source of Cellulase, Biotechnol Bioeng Symp No. 5:81

Matsumura M, Mark H (1986) Elimination of Ethanol Inhibition by Pervaporation, Biotechnol Bioeng 28:535

McCabe BJ, Eckenfelder WW Jr. (1955) Biological Treatment of Sewage and Industrial Wastes Reinhold, NY

Montenecourt B, Eveleigh D (1978) Proc Second Fuels from Biomass Symp, Rensselaer Polytechnic Institute, Troy, NY, 613–625

Montenecourt B, Eveleigh D (1977) Appl Environ Microbiol 34:777, 782

Mori, Y, Inaba T (1990) Biotechnol Bioeng 36:849

Mukerjee A, Lee YY, Tsao GT (1972) Gas-Liquid-Cell Oxygen Absorption in Fermentation, Fermentation Technology Today, Proceedings of the 4th International Fermentation Symposium, Kyoto, Japan, 65

Paiva TCB et al. (1996) Appl Biochem Biotechnol 535:57–58

Reese ET (1975) Biotechnol Bioeng Symp No. 5:71

Reese ET (1976) History of Cellulase Program at Natick, Biotechnol Bioeng Symp 6:9–30

Saddle J (1997) Steam Explosion of Softwood, presented at the 19th Symposium on Biotechnology for Fuels and Chemicals, Colorado Springs

Saeman J (1945) Ind Eng Chem 37:43

Schell D et al. (1998) Appl Biochem Biotechnol 17:70–72

Scholler H (1935) French Patent 777,824

Shoemaker SP et al. (1981) Trends in Biology of Fermentation, Plenum Press, pp 89–109

Shuler ML, Kargi F (1992) Bioprocess Engineering Prentice Hall, Englewood Cliffs, NJ

TAPPI (1988) Test Method T-222 om 88

Tsao GT (1968) Simultaneous Gas-Liquid Interfacial Oxygen Absorption and Biochemical Oxidation Biotechnol Bioeng 10:766

Tsao GT (1978) Cellulosic Materials as Renewable Resource, Process Biochem 13:12–14

Tsao GT, Gong CS, Cao NJ (1999) Repeated Solid Fermentation and Extraction for Enzyme Production, presented at the 21st Symposium on Biotechnology for Fuels and Chemicals, Fort Collins, Colorado, USA
Tsao GT, Ladisch MR, Voloch M, Bienkowski P (1982) Production of Ethanol and Chemicals from Cellulosic materials, Process Biochem 17:34–38
Underkofler LA, Hickey RJ (1954) Industrial Fermentation, vols 1 and 2. Chemical Publishing Commpany, NY
Whitworth DA et al. (1980) Ethanol from Wood, New Zealand Forest Services Report
Wilke CR (1975) editor, Cellulose as A Chemical and Energy Resource, Biotechnol Bioeng Symp No.5, Wiley, NY
Wilkinson RA (1996) *Zymomonas* for Ethanol Production, 11th International Symposium for Alcohol Fuels, Proceedings Vol. 2:379–95
Wu Z, Lee YY (1998) Nonisothermal SSF for Direct Conversion of Lignocellulosics into Ethanol, Appl Biochem Biotechnol 479:70–72
Wyman CE et al (1992) Ethanol and Methanol from Cellulosic Biomass, U N Solar Energy Group (SEGED) – Brazil, Proceedings, 865–923
Yange X, Tsai GJ, Tsao GT (1994) Enhancement of in situ Adsorption on Acetone-Butanol Fermentation by *Clostridium acetobutylicum*, Separation Technol 4:1–12
Yang X, Tsao GT (1995) Enhanced Acetone-butanol Fermentation Using Repeated Fed-Batch Operation Coupled with Cell Recycle by Membranes and Simultaneous Removal of Inhibitory Products by Adsorption, Biotechnol Bioeng 47:444–450
Yang X, Tsao GT (1994) Mathematical Modeling of Inhibition Kinetics in Acetone-Butanol Fermentation by *Clostridium acetobutylicum*, Biotechnol Progress 10:532–538
Zhang MQ et al. (1992) In-situ Separation of Ethanol Fermentation by CO_2 Stripping and Activated Carbon Adsorption Process, J Chem Ind Eng 7:19
Zheng Y et al. (1996) Avicel Hydrolysis by Cellulase Enzymes in High pressure Carbon Dioxide, Biotechnol Lett 18:451–454
Zheng Y et al. (1995) High Pressure Carbon Dioxide Explosion as a Pretreatment for Cellulose Hydrolysis, Biotechnol Lett 17:845–850
Zheng Y et al. (1996) Lactic Acid Fermentation and Adsorption on PVP, Appl Biochem Biotechnol 57–58, 627–632

Received July 1999

Advances in Enzyme Technology – UK Contributions

John M. Woodley

The Advanced Centre for Biochemical Engineering, Department of Biochemical Engineering, University College London, Torrington Place, London WC1E 7JE, UK

Enzyme technology has been a recognised part of bioprocess engineering since its inception in the 1950s and 1960s. In this article the early history of enzyme technology is discussed and the subsequent developments in enzyme isolation, enzyme modification and process technology are described. These creative developments have put enzyme technology in a position of huge potential to contribute to environmentally compatible and cost effective means of industrial chemical synthesis. Recent developments in protein modification to produce designer enzymes are leading a new wave of enzyme application.

Keywords. Enzyme isolation, Enzyme technology, Enzyme immobilisation, Protein engineering.

1 Introduction 94

2 Early History 94

3 Enzyme Production 95

3.1 Introduction 95
3.2 Cell production 96
3.3 Enzyme Isolation 97

4 Enzyme Immobilization 98

5 Bioprocess Technology 99

5.1 Introduction 99
5.2 Reactor Choice 100
5.3 Medium Choice 101
5.4 Process Integration 103
5.5 Process Design 103

6 Protein Engineering 104

7 Application and Future Perspectives 105

References 107

Advances in Biochemical Engineering/
Biotechnology, Vol. 70
Managing Editor: Th. Scheper

1 Introduction

Some 2500 enzymes have been identified to date [1] and currently around 250 are used commercially in various degrees of purity. However, only 25 enzymes account for 80% of all applications mainly in the processing of starch and for use as domestic and industrial detergent additives for cleaning clothes. The application of enzymes for industrial use is perhaps the widest definition of the term enzyme technology. The term has in the past 40 years been variously replaced by biocatalysis, bioconversion and biotransformation. Confusion has arisen where these latter terms may also be used to describe intact microbial cell catalysed reactions as well as isolated enzyme catalysed conversions. In this article I will focus on the use of enzymes (whether used in an intact cell or isolated) as potential catalysts for single step conversions and chart their development over the past 40 years. This has been a field of intense industrial and academic interest (from around 50 publications per year in the 1950s to a steady 750 per year in the 1990s [2]) and I will therefore necessarily be somewhat selective here.

2 Early History

While enzymes in various forms have been used for many thousands of years for the benefit of mankind, the application of enzyme technology to assist industrial and process chemistry is more recent. In the late nineteenth and early twentieth century there were a few isolated reports, but the first enzyme-based reactions of industrial importance were steroid modifications where the enzyme replaced a series of chemical steps required to introduce a hydroxyl group at a specific position [3]. Achieving such specificity is very difficult by conventional chemistry. The catalytic agent used was *Rhizopus nigricans* and the cells were grown in the presence of the reactant. This simple approach proved effective but widespread application was going to be limited to particular cases. For catalytic use it was recognised that ideally an isolated enzyme was required. Leaving aside the need to isolate the enzyme from the cells in the first place, it was also necessary to stabilize the enzyme (now that it was removed from the protective environment of the cell). One method to achieve this is to attach the enzyme to a solid support. Enzyme immobilization on, or in, a support not only provides a catalyst of sufficient size for ease of separation downstream of the reactor, but also keeps the enzyme structure rigid and therefore confers stability. The drive to achieve an immobilized enzyme began in the 1960s [4]. George Manecke described enzyme resins [5]. Malcolm Lilly [6] and Ephraim Katchalski [7] referred to water-insoluble enzyme derivatives and a fourth pioneer, Klaus Mosbach, described entrapped and matrix bound enzymes [8, 9]. However it was at a meeting in 1971 that the term immobilized enzyme was first used and became the standard nomenclature [10]. The 1971 meeting held at Hennniker in New Hampshire (USA) from August 9–13 became the first in a valuable series of Enzyme Engineering meetings held under the

Table 1. Advances in enzyme technology

Process	Advance
Growth process	High cell density growth
Enzyme isolation	Improved enzyme expression New enzyme isolation techniques Enzyme immobilization
Biotransformation	Protein engineering Use of non-aqueous media New reactor designs New reactor operating strategies
Catalyst recycle	Enzyme immobilization
Downstream processing	Integration of reaction with product recovery

auspices of the Engineering Foundation (now United Engineering Foundation). These biannual meetings, which are still running today, have formed the backbone of a strong international community in enzyme technology and we owe much to those early pioneers at that first meeting. The meetings aim to cross disciplinary boundaries and have served as an excellent forum for exchange of ideas. The subsequent years have seen a number of developments in enzyme production, enzyme modification and bioprocess technology. These have led to nine significant areas of advancement based on rDNA technology and process engineering research. These advances are listed in Table 1 and I will describe the development of each of these areas in this article.

3 Enzyme Production

3.1 Introduction

Despite early recognition of the need to isolate enzymes for use as biocatalysts, in a number of cases this is not the most effective catalyst form and several conversion types rely on operation in an intact cell (where the isolated enzyme is unstable or use is made of other endogenous enzyme activities (for example for cofactor recycle and regeneration)). There are three modes of operating an intact cell process: growing cell (where the cells are growing while conversion occurs), resting cell (where the cells are metabolically active but not growing while the conversion occurs) and resuspended resting cell (where the resting cells are resuspended in buffer to avoid the problems of product isolation from medium downstream from the reactor). The options are illustrated in Figure 1. Alternatively the enzyme may be isolated from the host organism prior to biotransformation, in order to reuse catalyst efficiently and/or reduce contaminating activities. This is particularly critical where enzymes are used as

Figure 1. Typical flowsheets for the three modes of intact cell-based biocatalysis

Growing cell	Resting cell	Resuspended resting cell
Cell production/ Biotransformation ↓ Cell recovery ↓ Product concentration ↓ Product isolation	Cell production ↓ Biotransformation ↓ Cell recovery ↓ Product concentration ↓ Product isolation	Cell production ↓ Cell recovery ↓ Resuspension ↓ Biotransformation ↓ Cell recovery ↓ Product concentration ↓ Product isolation

catalysts to assist in the synthesis of pharmaceuticals and final product purity will determine application. Regardless of the final biocatalyst form, the process begins with the production of the cell mass.

3.2 Cell production

A key development and significant reduction in the cost of biocatalyst production has come through the application of so called high cell density growth [11, 12] where bacteria can now be grown at large scale up to 100 g dry weight per litre. This is an order of magnitude improvement on previous biocatalyst production methods. Clearly it is necessary to control the growth rate of such processes to ensure that the oxygen demand is not too high, particularly with fast growing cells such as the commonly used strains of *Escherichia coli*. Routinely this is achieved through feeding the carbon source at a predefined rate to prevent oxygen demand outstripping supply. At large scale this has been found to be particularly important. Using such techniques, far higher titres of catalytic protein can be obtained from the cells. For those systems with well understood genetics a choice of host organism has also become possible on the basis of good expression or an operationally robust strain. Additionally the application of rDNA technology has now led to enzymes being induced more easily and cheaply (at large scale) and expressed at far higher levels in the cytoplasm, periplasmic space or even extracellularly. For instance at University College London we have cloned and overexpressed transketolase (for asymmetric carbon-carbon bond synthesis [13]) as high as 40% of the protein in the cell [14]. At such levels of expression inclusion bodies may be formed in some systems. However even when 15–20% of the protein is expressed as the desired enzyme, subsequent purification may be limited to removal of protease activity alone. Potentially such developments herald the advent of direct immobilisation from unpurified homogenised cells. One commercial process (operated by Glaxo

Wellcome in the UK) now uses this for the production of neuraminic acid aldolase for condensation of pyruvate with mannosamine to synthesise *N*-acetyl-D-neuraminic acid (sialic acid) as a precursor to the anti-flu compound, zanamavir [15].

3.2 Enzyme Isolation

In a limited number of cases enzymes are secreted into the growth medium, although at low concentration. However more frequently the useful enzyme activity is intracellular. Figure 2 is a schematic representation of a general flowsheet for an intracellular enzyme-based process. While a variety of enzyme sources is available, microbial routes are the most productive since the cells are quick to grow, genetics well understood (in some but not all cases), the cells can be broken and the debris separated from catalytic protein effectively. For these reasons most processes commence with a microbial growth process. In those cases where intracellular enzyme isolation is required (for example where transport into the cells is limited or there are competing enzyme-catalysed reactions) the growth process is followed by cell concentration prior to disruption and removal of cellular debris. Isolation of intracellular enzymes was intially done by chemical lysis. This was a difficult process and the ability to dissrupt cells via homogenisation [16] enabled cytoplasmic enzymes to be isolated for the first time [17]. This operation has been well characterised and is today standard practice for recovery of intracellular enzyme. The ability to clone into alternative hosts will also have importance here since some cells have been found to be much easier to break open than others [18]. Centrifugation to

Figure 2. Typical flowsheet for isolated intracellular enzyme-based biocatalytic processes

Cell production		
↓		
Concentration/Cell recovery	⟶	Water
↓		
Disruption		
↓		
Removal of cell debris	⟶	Cell debris
↓		
Immobilisation	⟵	Enzyme support
↓		
Biotransformation	⟵	Reactant
↓ ↑		
Catalyst recovery		
↓		
Concentration	⟶	Water
↓		
Product isolation	⟶	Side products
↓		
Product		

separate the cell debris, prior to purification has also become routine. Entrained air in homogenization and centrifugal operations is a particular problem since globular proteins are very susceptible to shear induced interfacial effects [19]. Surprisingly most enzymes have been found to be resistant to shear itself with the exception of some membrane bound enzymes [20]. Subsequent purification of the enzyme is required only to the extent of removing protease activity or contaminating activities that may affect the final yield of product. This is in sharp contrast to other biotechnological processes where protein purification is a dominant process issue. These techniques now mean that many enzymes can be recovered and isolated effectively for subsequent use as a suspended catalyst or immobilized onto a solid support.

4 Enzyme Immobilization

For intracellular isolated enzyme based catalysis the cost and difficulty of enzyme isolation means that the catalyst becomes a very significant part of the total cost. Hence there is a need to reuse and retain such enzymes. While this was the original justification for enzyme immobilization (necessary also with continuous reactors such as packed or fluidised beds [21]) this is only possible if the enzyme has sufficient stability. Fortunately in many cases the immobilization itself holds the structure of the protein in place and hence leads to significant improvements in the stability of the enzyme. In the 1960s the first work on 'insoluble enzymes' showed that using charged support materials the pH optimum and/or Michaelis constant could be shifted [22] for a particular enzyme. However these electrostatic effects, while academically interesting, did not lead to industrial development due to the high substrate concentrations required for commercial process operation. In 1969 the NSF in the USA began funding of enzyme engineering leading to a considerable interest in the development of immobilized systems. Table 2 lists some of the features of immobilised biocatalysts. One of the earliest applications was by the Tate and Lyle Sugar Company (in the UK) who operated a 6 metre deep bone char im-

Table 2. Features of immobilized biocatalysts

Process operation	Feature
Catalyst production	Loss of activity upon immobilization Additional cost of support
Reactor operation	Reduced activity per unit volume Diffusional limitations Increased stability Protection from interfacial damage[a] Options for alternative reactors[b]
Downstream	Improved separation

[a] Gas-liquid and liquid-liquid.
[b] Packed and fluidised beds.

mobilized invertase system in the 1940s. The academic research in the 1960s led to higher value processes such as the resolution of racemic mixtures of amino acids by Tanabe Seiyaku (in Japan) [23]. By 1978 at least seven different processes using immobilized biocatalysts were reported to be in commercial operation [24]. The biggest application, and that remains the case today, was for glucose isomerisation (GI) [25]. Since these enzymes are intracellular, justification of the isolation costs was only possible by enzyme multiple recycle. This was achieved via immobilization and used in a packed bed column with downward flow (to minimize residence time and avoid byproduct formation). Previous work had shown the kinetic benefits of operation in packed beds [26] and given very little pH change for the GI reaction an effective process was established. Today the process produces around 8 million tonnes per annum of high fructose glucose syrup. Another key reaction, which is one of the highest value immobilized enzyme processes, is for the cleavage of penicillin G (or V) using penicillin acylase to produce 6-aminopenicillanic acid (6-APA) as an intermediate in the production of semi-synthetic penicillins. Work between University College London and SmithKline Beecham (then Beecham) in the UK developed a process using isolated immobilized enzyme in a substrate fed batch reactor [27, 28]. This enabled pH adjustment to be carried out during the reaction to neutralize the formation of the byproduct, phenylacetic acid, and minimise the severe inhibition of enzyme activity by both products. Substrate inhibition was overcome by feeding. The process was introduced in 1973 and today around 7500 tonnes per annum of 6-APA are made using immobilized penicillin acylase [29]. The process has been much improved since its inception and now operates at an efficiency of about 2000 kg product/kg catalyst. Today many processes use immobilized enzymes and in a limited number of cases immobilized intact cells. In the last decade the use of cross linked enzyme crystals has been examined as an alternative to immobilization [30].

5 Bioprocess Technology

5.1 Introduction

While the power of enzyme catalysis for synthetic process chemistry was recognised in the 1950s, in many cases the properties of a specific biocatalyst were frequently far from ideal for industrial or process use. Since the late 1960s a number of innovative approaches have been applied to processes to overcome these features. Immobilization has been used to create water-insoluble enzyme particles for easy recovery and reuse, operation in organic solvent milieu has led to the effective conversion of poorly water-soluble compounds, reactor choice and judicious operating strategies have minimised reactant and product inhibition, and *in-situ* product removal has minimised product inhibition and toxicity and enabled thermodynamically unfavourable reactions to be carried out. These developments have been crucial to the area and enabled many reactions to be carried out commercially that would not have been possible otherwise. The techniques are relatively generic and in some cases rules have now been built to assist implementation [31].

5.2 Reactor Choice

Early steroid conversions were carried out with growing cells but it was soon recognised that for many applications the separation of the growth and conversion steps may be beneficial (Figure 1 second column)). Each individual part of the process may be optimised and product recovery effected from cleaner solutions without the presence of complex broth. Operation of a bioreactor and separate biocatalytic reactor also gives the opportunity for resuspension at different catalyst concentrations and use of alternative reactor designs. In other reactions operation with an isolated enzyme is preferable and here there are also choices concerning the optimal reactor configuration. In 1965 Lilly and coworkers [26] published a seminal work on the implications for reactor kinetics on operation in the three classical reactor configurations: batch stirred tank, continuous stirred tank and plug flow reactor. The analysis applied the approach of reaction engineers to evaluate reactors on a kinetic basis (ie which reactor made best use of the available enzyme activity) for enzymes following the Michaelis-Menten kinetic model. The results are presented in Table 3. There are clear advantages for operating in the batch and plug flow reactors with Michaelis-Menten kinetics, where high conversions are required and/or the Michaelis constant is high relative to the initial batch or feed substrate concentration [21, 22]. Another key consideration is the necessity for mixing (required where there is a second liquid phase present, oxygen present, solids present and/or pH control is a necessity) which in many cases will demand operation in a stirred tank configuration. The control of pH, required in any reaction where conversion leads to a change in pH via acid or base consumption or production, necessitates alkali or acid addition to neutralise the effects. Use of buffers is ruled out on the basis of cost, strength (given the concentrations of product required from stoichiometric conversions) and difficulties for

Table 3. Reactor kinetic analysis indicating best (●) use of available enzyme. Choice between batch stirred tank and packed bed is made on the basis of the need for mixing. If Michaelis constant is low and/or conversion required is low there is little difference between reactors

Catalyst kinetics	Reactors		
	Stirred tank		Plug flow
	Batch	Continuous	Packed bed
Michaelis constant high/ High conversion	●	○	●
Substrate inhibition	○	●	○
Product inhibition	●	○	●

downstream product recovery [31]. Consequently most reactors used industrially are batch stirred tanks or fed-batch versions to reduce substrate inhibition. However the introduction of immobilized enzyme as a catalyst form raised the possibility of operation in a plug flow reactor (as a packed bed providing the beads were incompressible). This reactor type not only has kinetic advantages (see Table 3) but also enables a far higher concentration of catalyst within the reactor (voidage around 34% compared to 90% in a stirred tank reactor on account of attrition). For some reactions there has been a clear advantage operating in this way (e.g. fat interesterification and glucose isomerization). In the 1970s a number of reports appeared on the use of fluidised bed reactors [33]. Fluidised beds offer the advantage of high catalyst loading with the possibility of operating with a second phase (liquid, solid or gas). Dependent on the extent of bed expansion, operation can be in plug flow or well mixed regimes. More recently magnetically stabilised beds have been examined as an option to reduce the bead size in a fluidised bed (at the same flowrate) to effect conversions with minimal diffusional limitations [34]. Membrane reactors and a number of other novel configurations have also been examined since the 1980s for multiphasic conversions (and cofactor recycle systems) [35] although commercial application has been limited.

5.3 Medium Choice

While perceived wisdom holds that enzymes operate in water, as early as the nineteenth century there were reports of enzyme catalysed reactions carried out in organic solvents [36,37]. However it was only in the 1970s that the potential of this for enzyme technology was recognised. An isolated report in 1936 [38] was followed by pioneering work in the 1970s by Carrea (in Italy) [39] and Lilly (in the UK) [40]. They argued that the majority of useful compounds for organic chemists are poorly water-soluble and hence, in aqueous-based reaction media, are present at very low concentrations. They introduced water immiscible organic solvents to improve the process by changing the reaction medium [41–43]. Aside from kinetic implications the real benefit lies in higher product concentrations to assist downstream processing. In many cases the product was present in the organic solvent which also assisted isolation. Early experiments involved subjecting cell and enzyme systems to aqueous media which were replaced by organic solvent in increasing proportions.

Klibanov and coworkers [44, 45] took this concept to the extreme by studying the role of water on enzyme catalysis and found that only a fraction of the protein surface need be covered with water for it to remain active. Remarkably this ability to operate in near neat organic solvent led to the possibility of running hydrolytic reactions in reverse (esterification) or even in solvent without water taking part in the reaction (transesterification, interesterification) and a number of processes have now been commercialised. Unilever saw potential in the interesterification reaction for upgrading fats converting mid palm fraction into a substitute cocoa butter and in 1976 described a process using an immobilized enzyme for fat interesterification [46]. The substrate was fed to a packed bed reactor containing hydrated celite as the support for the lipase. More advanced particles have been introduced subsequently. Maintaining the correct amount of water is critical since too low a level results in a reduced activity and too high a level leads to fat hydrolysis (rather than interesterification) giving a poor quality product containing diglycerides. This reaction exploited the unique properties of lipases which operate effectively at an organic-aqueous liquid-liquid interface. Other enzymes have been shown to work in the bulk of the aqueous phase. However in all cases where organic solvents were introduced the effects were marked, frequently giving increased productivities and better access of the poorly water soluble substrates to the biocatalyst. In the 1980s limitations were also identified for process application. In all but the most insoluble solvents the biocatalyst was denatured to some extent and the amount of water present found to be very critical. An important line of research was commenced examining the role of water on activity [47], the nature of the organic solvent [48,49] and the amount of organic solvent present [50]. Three types of system can be identified: Those in a homogeneous aqueous medium (with water-miscible organic solvent present); those in an homogeneous organic medium (with a small amount of water dissolved in water-immiscible organic solvents) and heterogeneous two-liquid phase systems (water immiscible organic solvent and an aqueous phase forming a separate phase). Each system has found industrial application, although there are particular process reasons for use of a two-liquid phase system [51]. Some representative examples of multiphasic systems are given in Table 4. While lipases function well at an interface and at low water activities, whole cell catalysts require enough water to form a second phase (activity greater than unity). Likewise water may be the substrate or the product of the reaction itself or there may be a shift in relative water solubilities from the substrate to the product. Oxidations are particularly good examples of this where the insertion of the oxygen leads to often dramatic increases in the water solubility of the compound. While in reaction examples such as fat interesterification the use of an organic solvent in the process is a prerequisite for activity, in some cases the addition of an organic solvent may provide an alternative process option. This is well illustrated with a process developed by ICI (now Avecia Life Science Molecules) for the production of *cis*-1,2-dihdro-1,2-dihdroxy-cyclohexa-3,5-diene (DHCD). DHCD may be polymerised to form a precursor to polyphenylene. Polyphenylene has many unique properties and the synthesis of DHCD is chemically intractable. ICI developed

Table 4. Representative examples of multiphasic biocatalytic systems. Where LCA is long chain alkane such as *n*-hexadecane or tetradecane

Aqueous	Medium Organic	Gas	Reaction	Biocatalyst
Water	LCA	Air	Alkene epoxidation	Intact cell
	LCA	Interesterification	Immobilized lipase	
Water	Reactant	Ester hydrolysis	Lipase/esterase	
Water	LCA	Air	Aromatic hydroxylation	Intact cell

a process using a blocked mutant of *Pseudomonas putida* to hydroxylate benzene into DHCD. The reaction was first demonstrated by Gibson and coworkers [52] and the process for biocatalytic conversion first described by Taylor in 1975. While the *cis* glycol product is water soluble the aromatic substrate is only soluble to around 0.5 g/l in water. The substrate can be fed to a batch reactor (although control is difficult) [53] and an alternative is to use a second organic phase. Experiments with addition of tetradecane to form a two-liquid phase system proved particularly successful resulting in up to 60 g/l of product [54]. While the use of organic solvents in biocatalytic reaction media is now finding wide acceptance, debate over the optimal choice of solvent for a particular reaction continues [55, 56]. More recently biocatalysis has been carried out effectively in slurries (with solid substrate and/or product present) and in gas phase media.

5.4 Process Integration

Many biocatalytic reactions are productivity limited by the presence of the product. As the product concentration builds in the reactor it may inhibit the enzyme activity or be toxic to the cell. Likewise reactions which are not thermodynamically favourable may be improved by removal of the product as the reaction proceeds to shift the equilibrium in the synthetic direction. Theoretically such conversions could also be improved by operation with an excess of substrate but this is often precluded on the grounds the substrate is inhibitory to the enzyme, thereby compromising kinetics. *In-situ* product removal (ISPR), where the product is removed from the reactor as it is formed, may therefore provide a tool to overcome some of these problems. Until the 1980s work had focussed in large part on the extractive removal of organic acids and organic solvents from growth processes [57]. More recently extractive ISPR has been examined in relation to the products of enzyme-catalysed reactions [58]. While extraction into an organic solvent gives good capacities for the product being removed, it is not particularly selective. The key need in a biocatalytic reaction is to separate the product from a structurally related substrate. This therefore requires particular selectivity to achieve effective removal [59]. However a number of industrial processes have now used this approach to enable process implementation [60].

5.5 Process Design

By the mid 1980s it became clear that many enzyme based reactions could be implemented for commercial operation provided one of the growing number of process technology options could be invoked. Woodley and Lilly [31, 61] have pursued the idea of a systematic approach to process choice and selection so as to match the process with the biocatalyst features. An approach was devised based on the simple premise of a relationship between reaction and biocatalyst properties with the final process. This systematic approach (schematically il-

Figure 3. Systematic approach to biocatalytic process selection

Biocatalyst/Reaction properties
↓
Process constraints
↓
Eliminate poor options
↓
Detailed optimisation (limited cases)

lustrated in Figure 3) enables property data to define constraints which then guide process decision making. In essence the approach reduces the number of options for detailed evaluation by eliminating those that are ineffective early on. Part of the rationale for this research was the need to make process design decisions quickly, especially in the pharmaceutical industry where pressure to get to market is great. More recently the representation of the process data has been examined in the form of "process maps" – a graphical technique to visualise the conditions acceptable to the biocatalyst in the reactor [62, 63]. This enables sensitivity analyses and process flexibility along with alterations to the process to be examined using a graphical tool. Combined with high throughput techniques to collect the data to build the map, this will further accelerate the decision making process.

6 Protein Engineering

Most enzymes in nature operate in a highly evolved environment. Unlike chemical catalysts, biological catalysts have evolved over many thousands of years to operate in a narrow range of pH, temperature, ionic strength and concentration. In nature therefore enzymes operate very effectively. However for industrial catalysis three problems are presented. First the enzymes operate at high rates of conversion on natural substrates with the consequence that only low rates are observed with non-natural substrates. This means that conversion of the most interesting, novel, compounds is slow. The second problem concerns the limited range of substrate and product concentrations which give good enzymatic rates and biocatalyst stability. The necessity of efficient downstream product recovery from what is usually an aqueous based medium requires reactor leaving concentrations to be 20–60 g/l for a fine chemical product and an even higher concentrations for lower value industrial (bulk) chemicals. The majority of reactions in nature operate at a mg/l level and hence downstream processing becomes difficult. Thirdly many enzymes operate in a narrow window of pH and temperature. This may not be the same window defined by the substrate and product of a specific reaction. The resulting compromise costs catalytic efficiency and conversion yield. While process techniques can be used to overcome a number of biocatalyst productivity limits, knowledge of enzyme structure (as a result of protein crystallisation developments) in the 1980s

Figure 4. General strategy for directed evolution of enzymic catalysts

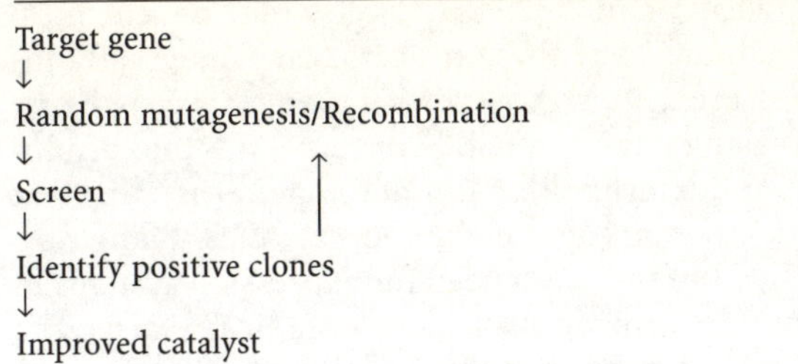

raised the intriguing possibility of altering amino acid residues around the active site in order to change the catalyst properties. However the relationship between amino acid changes and the resultant catalytic properties have proved in many cases rather elusive. More recently in the 1990s another approach has been taken. By making alterations in the gene by random mutagenesis (rather than site specific) and/or recombination techniques huge diversity can be generated. This can now be ordered using developments in high throughput screening for favourable properties. Screening these mutants enables selection of positive clones which can be fed back to the mutagenesis stage. In this way the laboratories of Arnold and Stemmer have been able to mimic evolution [64, 65], as illustrated in Figure 4 and preferentially alter biocatalytic properties for industrial exploitation [66]. The work is still in its infancy but the results are impressive, with order of magnitude improvements in specificity, activity and stability in a range of process conditions. The techniques rely on an effective screen and ability to express the genes in a suitable host. Thus far relatively simple proteins have been improved and the goal is to develop appropriate protocols for more complex targets such as redox enzymes [67, 68].

7 Application and Future Perspectives

As discussed earlier the first attempts at commercialisation of enzyme catalysed reactions were in the 1950s and 1960s. It is widely believed that today there are over 100 commercially operating processes from diverse companies. The majority of reactions are in the hydrolytic field primarily with the objective of racemic resolution. This a proven technology for these applications. More interesting however are the academic developments in asymmetric carbon-carbon bond formation and redox catalysis which are now starting to be commercialised and which will change the chemistries to which biocatalysis contributes over the next decade [69–71]. While multistep conversions will see application with intact cell biocatalyst format, for some cofactor-dependent single step conversions the choice between intact cell and isolated enzyme (with cofactor recycle) is becoming a real possibility [72]. The catalyst choice for such conversions will become a particularly interesting area of research in the coming years. Enzyme technology will find a key role in the synthesis of complex pharmaceuticals and carbohydrates, in particular where optically pure products are required. Their application in the field of large scale bulk chemical synthesis is harder since the economics of these processes means greatly increased product concentrations are required. Nevertheless headway is being made here too and given increasing environmental concerns, the role of biocatalysis is set to increase. While the technology has developed it is clear that there is a need for a greater source of commercialised enzymes. Many new activities are now being identified but there is a need for expression systems in suitable hosts and stable enzymes to be made available to organic chemists. The ability to modify the catalyst and create improved biocatalysts as well as overexpress the enzyme, together with effective growth and isolation techniques will be increasingly required by companies in the field.

Enzyme technology has been a field of intense research activity over the past 40 years and it is instructive to note a number of trends in the context of this historical review which point to directions for the future.

- First the range of reactions to which enzyme catalysis can be applied is very broad although currently this is not being exploited. Academic research into carbon-carbon bond formation and redox based reactions will increasingly find industrial application. It is vital that biologists work closely with organic chemists to identify targets.
- Secondly the relative costs which make up the economic analysis of the products of these reactions is changing. Early work was rightly very concerned with yield of product on the catalyst. The catalysts were expensive and not readily available. Since the ability to clone and overexpress enzyme activity, the relative cost of the catalyst has reduced. Alongside this the value of the substrates has increased. Ever more complex substrates with multichiral centres are now the starting points for biocatalysis. Hence yield on reactant is becoming increasingly critical. This changes the process analysis and the appropriate technology to enable implementation.
- Thirdly there has been much progress in the implementation of process techniques to overcome non ideal biocatalyst properties. More recently it has become possible to alter the properties of the biocatalyst via genetic and protein engineering. The combination of these approaches will be very powerful and lead to implementation of completely new groups of biocatalysts.

These trends give a perspective on biocatalysis which indicate a bright future providing chemists, biologists and biochemical engineers continue to work effectively together in the way they have come to learn from enzyme technology in the past few decades.

Acknowledgement. Malcolm Lilly, who was to have authored this chapter, sadly died in May 1998. It is therefore dedicated to him as my colleague and friend, whose inspiration and intellectual leadership was in so large a part responsible for the development of enzyme technology.

References

1. Faber K (1995) biotransformations in Organic Chemistry, Springer, Berlin Heidelberg New York
2. Meyer H-P, Kiener A, Imwinkelvied R, Shaw N (1997) Chimia 51:287
3. Peterson DH, Murray HC, Eppstein SH, Reineke LM, Weintaub A, Meister PD, Leigh HM (1952) J Am Chem Soc 74:5933
4. Kay G (1968) Process Biochem 3(8):36
5. Manecke G (1962) Pure Appl Chem 4:507
6. Lilly MD, Money C, Hornby, WE, Crook EM (1965) Biochem J 95:45
7. Bar-Eli A, Katchalski E (1963) J Biol Chem 238:1690
8. Mosbach K, Mosbach R (1966) Acta Chem Scand 20:2807
9. Mosbach K (1970) Acta Chem Scand 24:2084
10. Sundaram PV, Pye EK, Chang TMS, Edwards VH, Humphrey AE, Kaplan NO, Katchalski E, Levin Y, Lilly MD, Manecke G, Mosbach K, Patchornik A, Porath J, Weethall HH, Wingard LB (1972) Biotechnol Bioeng Symp 3:15
11. Reisenberg D, Menzel K, Schulz V, Scumann K, Vieth G, Zuber G, Knorre AW (1990) Appl Microbiol Biotechnol 34:77
12. Reisenberg D, Schulz V, Knorre WA, Pohl HD, Karz D, Sanders EA, Ross A, Deckwer WD (1991) J Biotechnol 20:17
13. Hobbs GR, Lilly MD, Turner NJ, Ward JM, Willetts AJ, Woodley JM (1993) J Chem Soc Perkin Trans I(2):165
14. Lilly MD, Chauhan R, French C, Gyamerah M, Hobbs GR, Humphrey A, Isupov M, Littlechild JA, Mitra RK, Morris KG, Rupprecht M, Turner NJ, Ward JM, Willetts AJ, Woodley JM (1996) Ann NY Acad Sci 782:513
15. Mahmoudian M, Noble D, Drake CS, Middleton RF, Montgomery DS, Piercey JE, Ramlakhan D, Todd M, Dawson MJ (1997) Enzyme Microb Technol 20:393
16. Hetherington PJ, Follows M, Dunnill P, Lilly MD (1971) Trans Instn Chem Engrs 49:142
17. Dunnill P, Lilly MD (1972) Biotechnol Bioeng Symp 3:97
18. Buckland BC, Richmond W, Dunnill P, Lilly MD (1974) The large-scale isolation of intracellular microbial enzymes: cholesterol oxidase from *Nocardia*. In: Spencer B (ed) Industrial Aspects of Biochemistry. FEBS, Amsterdam, p 65
19. Thomas CR, Dunnill P (1979) Biotechnol Bioeng 21:2279
20. Talboys BL, Dunnill P (1985) Biotechnol Bioeng 27:1730
21. Lilly MD, Dunnill P (1971) Process Biochem 6(8):29
22. Goldstein L,Levin Y, Katchalski E (1964) Biochemistry 3:1913
23. Tosa T, Mori T, Fuse N, Chibata I (1969) Agr Biol Chem 33:1047
24. Lilly MD (1979) Dechema Monograph 82:165
25. Lilly MD (1994) Chem Eng Sci 49:151
26. Lilly MD, Sharp AK (1968) The Chemical Engineer 215:CE12
27. Self DA, Kay G, Lilly MD (1969) Biotechnol Bioeng 11:283
28. Warburton D, Dunnill P, Lilly MD (1973) Biotechnol Bioeng 15:13
29. Savidge TA (1984) Enzymatic conversions used in the production of penicillins and cephalosporins. In: Vandamme E (ed) Biotechnology of Industrial Antibiotics. Marcel Dekker, New York, p 171
30. Tischer W, Kasche V (1999) Trends Biotechnol 17:326
31. Woodley JM, Lilly MD (1994) Biotransformation reactor selection and operation. In: Cabral JMS, Best D, Boross L, Tramper J (eds) Applied Biocatalysis. Harwood Academic, Chur, p 371
32. Lilly MD, Dunnill P (1972) Biotech Bioeng Symp 3:221
33. Lieberman RB, Ollis DF (1975) Biotechnol Bioeng 17:1401
34. Sada E, Katoh S, Shioza M, Fukui T (1981) Biotechnol Bioeng 23:2561
35. Cabral JMS, Tramper J (1984) Bioreactor design. In: Cabral JMS, Best D, Boross L, Tramper J (eds) Applied Biocatalysis, Harwood Academic, Chur, p 333
36. Hill A C (1898) J Chem Soc 73:634

37. Kastle JC, Loenhart AS (1900) Am Chem J 24:491
38. Sym EA (1936) Biochem J 30:609
39. Cremonesi P, Carrea G, Ferrara L, Antonini E (1975) Biotechnol Bioeng 17:1101
40. Buckland BC, Dunnill P, Lilly MD (1975) Biotechnol Bioeng 17:815
41. Lilly MD (1982) J Chem Tech Biotechnol 32:162
42. Lilly MD (1983) Phil Trans R Soc London 300:391
43. Carrea G (1984) Trends Biotechnol 2:102
44. Zaks A, Klibanov AM (1984) Science 224:1249
45. Zaks A, Klibanov AM (1985) Proc Natl Acad Sci USA 82:3192
46. Macrae AR (1983) J Am Oil Chem Soc 60:291
47. Halling PJ, Valivety RH (1992) Progress Biotechnol 8:13
48. Brink LES, Tramper J (1985) Biotechnol Bioeng 27:1258
49. Laane C, Boeren S, Vos K (1985) Trends Biotechnol 3:251
50. Lilly MD, Dervakos G, Woodley JM (1990) Two-liquid phase biocatalysis: choice of phase ratio. In: Copping LG, Martin RE, Pickett JE, Bucke C, Bunch AW (eds) Opportunities in Biotransformation, Elsevier, London, p 5
51. Woodley JM, Lilly MD (1992) Progress Biotechnol 8:147
52. Gibson DT, Hensley M, Yoshoka H, Mabry TJ (1970) Biochemistry 9:1626
53. Taylor SC, Brown S (1986) Perf Chem :20
54. Collins AM, Woodley JM, Liddell JM (1995) J Ind Microbiol 14:382
55. Schmid A, Kollmer A, Mathys RG, Witholt B (1998) Extremophiles 2:249
56. Leon R, Fernandes P, Pinheiro HM, Cabral JMS (1998) Enzyme Microb Technol 23:483
57. Roffler SR, Blanch HW, Wilke CR (1984) Trends Biotechnol 2:129
58. Freeman A, Woodley JM, Lilly MD (1993) Bio/Technology 11:1007
59. Lye GJ, Woodley JM (1999) Trends Biotechnol 17:395
60. Vicenzi JT, Zmijewski MJ, Reinhard MR, Landen BE, Muth WL, Marler PG (1997) Enzyme Microb Technol 20:494
61. Lilly MD, Woodley JM (1996) J Ind Microbiol 17:24
62. Woodley JM, Titchener-Hooker NJ (1996) Bioprocess Engng 14:263
63. Blayer S, Woodley JM, Lilly MD, Dawson MJ (1996) Biotechnol Progr 12:758
64. Affholter J, Arnold FH (1999) Chemtech 29:34
65. Crameri A, Raillard SA, Bermudez E, Stemmer WP (1998) Nature 391:288
66. Zhao H, Moore JC, Volkov AA, Arnold FH (1999) Methods for optimizing industrial enzymes by directed evolution. In: Demain AL, Davies JE (eds) Manual of Industrial Microbiology and Biotechnology, 2nd edn. ASM Press, Washington DC, p 597
67. Joo H, Lin Z, Arnold FH (1999) Nature 399:670
68. Roberts GCK (1999) Chem Biol 6:R269
69. McCoy M (1999) Chen Eng News 77:10
70. Drauz K, Waldmann H (eds) (1995) Handbook of Enzyme Catalysis in Organic Synthesis. VCH, Weinheim
71. Wong C-H, Halcomb RL, Ichikawa Y, Kajimoto T (1995) Angew Chem Int Ed 34:421
72. Kula M-R, Wandrey C (1987) Meth Enzymol 136:9

Received June 1999

Computer Applications in Bioprocessing

Henry R. Bungay

Howard P. Isermann, Department of Chemical Engineering, Rensselaer Polytechnic Institute, Troy, NY 12180-3590, USA
E-mail: bungah@rpi.edu

Biotechnologists have stayed at the forefront for practical applications for computing. As hardware and software for computing have evolved, the latest advances have found eager users in the area of bioprocessing. Accomplishments and their significance can be appreciated by tracing the history and the interplay between the computing tools and the problems that have been solved in bioprocessing.

Keywords. Computers, Bioprocessing, Artificial intelligence, Control, Models, Education.

1 Introduction 110

2 Historical Development 111

3 Biotechnology 117

3.1 Simulation 117
3.2 Monitoring and Control of Bioprocesses 119
3.3 Bioprocess Analysis and Design 120

4 Recent Activities 120

4.1 Models 120
4.1.1 Unstructured Models 121
4.1.2 Structured Models 121
4.2 Bioprocess Control and Automation 122
4.2.1 Sensors 123
4.2.2 Observers 123
4.2.3 Auxostats 124
4.2.4 Examples 124
4.2.5 Modeling and Control of Downstream Processing 125
4.3 Intelligent Systems 126
4.3.1 Expert Systems 127
4.3.2 Fuzzy Logic 127
4.3.3 Neural Networks 128
4.4 Responses of Microbial Process 129
4.5 Metabolic Engineering 130

5 Information Management 130

5.1 Customer Service 131
5.2 Electronic Communication and Teaching with Computers 131

Advances in Biochemical Engineering/
Biotechnology, Vol. 70
Managing Editor: Th. Scheper

6 **Some Personal Tips** 133

7 **Conclusions and Predictions** 134

Appendix: Terminology for Process Dynamics and Control 135

References 137

1 Introduction

To provide some historical perspective about what people were doing with computers and what has changed, I will follow the personalized approach used by others [1]. While pursuing my B. Chem. Eng. and Ph. D. degrees in the late 1940s and early 1950s, I had no contact at all with computers. My thesis was typewritten with carbon copies. After working for more than 7 years at a large pharmaceutical firm where the technical people thought that computers were for payrolls and finance and not of much use for research and development, I joined the faculty of a university in 1963 where about 20% of the engineering professors worked with computers. My education in chemical engineering was not current because my Ph. D. was in biochemistry. I audited a series of five courses in mathematics, studied process dynamics, helped teach it, and thus upgraded my engineering skills.

It was obvious that engineers who used computers could compete better in the real world, so I sought ways to apply computing in both teaching and research. Some professors still rely on their students for any computing, but I felt then and continue to think that you cannot appreciate fully what computers can do when you cannot write programs. I learned FORTRAN but regressed to BASIC when I began to work mostly with small computers. Along the way I have written a few Pascal programs and have dabbled with languages such as Forth. Early in 1997, I switched to Java which presented a very steep learning curve for me because of its object orientation.

I left teaching for another stint in industry from 1973 until 1976. I was in management and ordered a minicomputer for my technical staff. I was the person who used it most but for fairly easy tasks. One program that solved a production problem was for blending of a selection of input lots of stale blood to get adequate values of different blood factors in a product used for standardizing assays in a hospital laboratory. I became fully comfortable with a minicomputer, but my level of sophistication of programming changed little. Our only project related to getting computers into manufacturing tried electronic data logging at the process and carrying the records to the computer for analysis [2].

By the time I returned to teaching, minicomputers were common. The mainframe computer was widely used, but we also had rooms full of smart terminals that were fed their programs from a server. Very soon my research required a computer in the laboratory because we focused on dynamics and control. For over 20 years we have improved our systems incrementally by upgrading and extending both our hardware and software. All of my graduate students have studied process control, and most have used it in their research. Interfacing a bioreactor to a computer is routine for us, and some of our control algorithms are quite sophisticated. We make some use of artificial intelligence.

2 Historical Development

When I entered academia, analog computers were important. We think of the high-speed of digital computers, but analog computers are lightning fast when handling systems of equations because their components are arranged in parallel. They integrate by charging a capacitor. With large capacitors, voltages change slowly, and the output can be sent to a strip chart recorder or X-Y plotter. Small capacitors give rapid changes with the results displayed on an oscilloscope. Each coefficient is set with a potentiometer, and the knobs can be twisted for testing coefficients while watching the graphs change. This used to be far more convenient than making runs with a digital computer that had essentially no graphical output; the digital results had to be compared as columns of numbers on printed pages. Analog computers have about the same precision as a slide rule, but we are spoiled by the many figures (often insignificant) provided by a digital computer. The Achilles heel of the analog computer is the wiring. Each differential equation requires an integrating circuit; terms in the equation are summed at the integrator's input. Voltages are multiplied by constants by using potentiometers. Constants are developed by taking a fraction of a reference voltage, either plus or minus. With many components, jacks for reference voltages, wires going everywhere for interconnections, jacks for inputs and outputs to pots, jacks for initial conditions, and the like, the hookup for a practical problem resembles a rat's nest. Furthermore, a special unit is needed for each multiplication or division, and function generators handle such things as trig relationships and logarithms. To summarize, analog computers perform summation, integration, and multiplication by a constant very well but are clumsy for multiplication or division of two variables and for functional relationships.

Scaling could sometimes be a chore when setting up an analog computer circuit. The inaccuracy can be great when a constant is not a significant fraction of a reference voltage. Consider, for example, the constant 0.001 to be developed from a reference voltage of 10 V. The pot would have to be turned to almost the end of its range. Proper technique is to scale the constant up at this point and to scale its effect back down at a later point. In addition to magnitude scaling, there can be time scaling when rate coefficients are badly matched.

I spent a fair amount of time with analog computers and enjoyed them very much. I used them for teaching because students could watch graphs change as they tested permutations of coefficients. One terrible frustration with the computers that were used for instruction was bad wires. Students, although admonished not to do so, were thoughtless in yanking wires out of a connection. The wires would come apart inside the plugs where the fault was not visible. Debugging a huge wiring layout and finding out hours later that one or more of the wires was broken could ruin your day.

I did a little hybrid computing after learning how to do so in a manufacturer's short course. The concept is to let a digital computer control an analog computer. The example most quoted for using a hybrid computer was calculations for a space vehicle. The digital computer was better for calculating the orbit or location and the analog computer, with its parallel and fast interplay, was better for calculating pitch, yaw, and roll. The messy wiring and the difficulty of scaling voltages to match the ranges of the variables doomed both analog and hybrid computation to near extinction soon after digital computers had good graphical output.

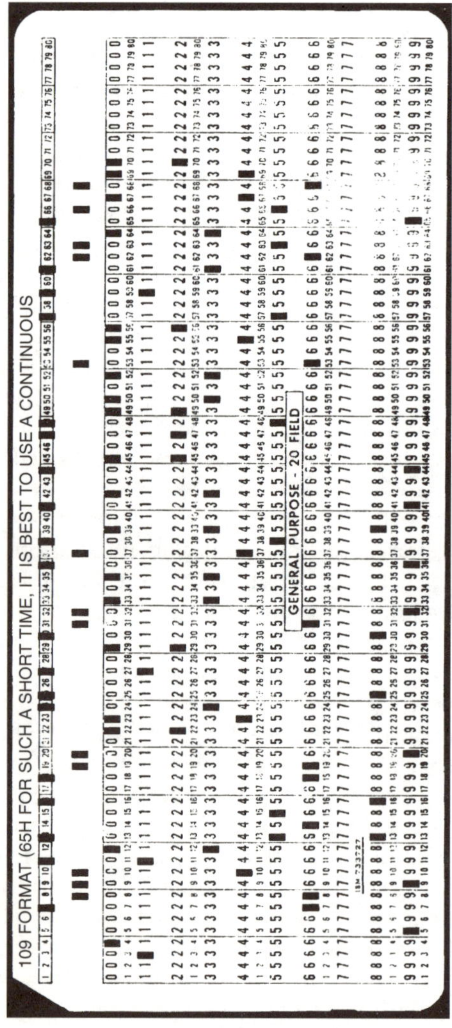

Fig. 1

In the early 1960s, FORTRAN was the most popular language for engineers by far. I learned FORTRAN from books and by examining programs written by others and began to integrate some digital computing into my courses. There were several companies that manufactured main frame computers, and FORTRAN code that I wrote at my university required some modifications before it could execute on another system when I spent the summer of 1970 at a different university.

The IBM punch card was used for communicating with the computer. A typical punched card is shown as Fig. 1. An entire, deep box of cards might be needed to feed the program and the data into the computer. Typical turn around time was overnight, and long runs might not be scheduled for two or three days. Many people were delighted when computer centers could furnish results in an hour or two. Today we have rooms full of personal computers or work stations. In the mid 1960s and through the early 1970s there were rooms full of noisy IBM machines for punching cards. These were fed into a card reader. Wide paper fed on rolls to the printer ended up fan folded with your results. You separated pages along the perforations and held them in thick books with metal strips passed through holes in the paper. There was no graphic output from the printer except when you devised a way to arrange characters as a crude graph. To get real graphs you requested a line printer where a pen moved across the page and touched the paper to make points or lines as the paper was moved back and forth underneath.

Despite the primitive equipment, much could be done. Libraries of code were available for various routine tasks such as a least squares fit of an equation to data points. Remember that the pocket calculator was not common until about 1970 and that mechanical calculators were big, clumsy, noisy, and not very powerful. Feeding punch cards to a computer seemed the best way to calculate even when answers were not ready for a few hours. You could get decks of cards for statistical routines and for various engineering calculations, attach your data cards, feed the whole pile into a card reader, and return later to the computer center for your printouts, often far into the night when you were trying for as many runs as possible. The programs that I wrote were mostly for numerical solutions of equations. I devised a game that taught my students in biochemical engineering a little about bioprocess development [3]. The punch cards had 72 spaces (fields), so I decided upon 7 variables (sugar concentration, amount of oil, percentage of inoculum, etc.) that each took 10 spaces.

The minicomputer caused a revolution in attitudes. For the first time, the ordinary user could sit at the computer and work interactively with programs. Paper tape replaced punch cards, and magnetic storage devices soon took over. Digital Equipment Corporation sold minicomputers such as their PDP-8 that was inexpensive enough for a few people to share. There was one just down the hall from my office, and I could use it for 4 or 5 h each week. Memory was limited, and programming was at the processor level. You had to code each operation. For example, multiplication required moving binary numbers in and out of the central processor, shifting bits, and keeping track of memory locations. Working with floating point numbers with some bits for the characteristic and others for the mantissa was not easy. You learned to think in binary and then in octal because it was less cumbersome. Before long there were languages that could simplify operations at an assembly level. Just about the time I learned one, higher level minicomputer languages appeared soon to be followed by compilers for real languages such as FORTRAN. Now you could write code easily, debug interactively, and perform what-if experiments with your programs. Coils of paper tape for storing programs were superseded by flat-fold paper tape. Very tough plastic tape was used to some extent.

Minicomputers made it practicable to dedicate a computer to a process. Groups such as that led by Humphrey at the University of Pennsylvania developed ways to interface a computer to a bioreactor. Numerous students wrote new code or improved the code of other students. Much was learned about sensors, signal conditioning, data display, and process analysis. The concepts were the bases for commercial software, but the code from the early days is mostly obsolete. That is not to say that some groups do not still write code for computer interfacing, but chances are that commercial software will handle most tasks [4]. Instead of a year or more for writing your own program, learning to use commercial software takes perhaps 2–6 weeks.

Personal computers intruded on the monopoly of minicomputers, and you could own a computer instead of sharing with others. The first magnetic storage that was affordable was an audio tape cassette recorder; the stream of bits from the computer produced sounds that could be played back and reconverted to bits. A program might be saved as three or four different files to have high probability that at least one copy would function properly.

My first personal computer, an Altaire, was build from a kit in 1976 and had 12 kilobytes of memory. A short program had to be toggled in with switches on the console before the computer could read from a paper tape. You tended to leave your computer on overnight because mistakes were common when toggling, and it could be highly annoying to get it booted again. The version of BASIC that I used took more than 8 kilobytes of the 8-bit memory, leaving little for the code written by me. One inexpensive way to add memory 4 kilobytes at a time was to wire a kit for a circuit board, insert memory chips, and plug the board into the computer.

I must express deep gratitude to students who worked part-time in my laboratory. We usually had a student from electrical engineering who could build devices and troubleshoot problems. Today, all of us can be frustrated when installing new hardware or a new program because the instructions are not always clear and because following the instructions is no guarantee that the results will be satisfactory. This is a picnic compared to debugging problems in the early days. With our home-built computers it was essential to trace circuits, identify bad chips, and to match cables to the ports. When we had better PCs, these electrical engineering students were still of great value for constructing sensor circuits, matching impedances, fixing the A/D converters, connecting stepping motors, and the like. We built our own preamplifiers for $10 worth of parts, and they performed as well as units costing between $500 and $1000. My students complained about taking time to construct and test electronic circuits, but I met students at other universities who complained about equivalent electronic devices that they purchased. There are delays in shipping and lost time for service with commercial equipment. When something went wrong with a home-made circuit, we fixed it in a matter of hours instead of waiting for days or weeks to get outside service. My students learned enough simple electronics to impress the other graduate students in chemical engineering.

An early input/output device was the teletype. It combined a typewriter, printer, and paper tape punch/reader. Service with a computer was demanding, and repairs were frequent. I recall being responsible for three primitive PCs that were used by students. Each had a teletype, and few weeks went by without lugging one teletype out to my car and going off to get it fixed. Dot matrix printers made the teletype obsolete. These first printers were noisy, and enclosures to deaden their sound were popular. Cost of a printer for your PC approached $1000, and performance was much inferior to units that cost $150 today. I have owned dot matrix printers, a dot matrix printer with colored ribbons, a laser printer, and most recently an ink jet color printer that eats up ink cartridges too quickly.

My next personal computer was similar to the Altaire, but with read-only memory to get it booted and an eight-inch floppy disk drive. There was some software for crude word processing. Much of the good software came from amateurs and was distributed by computer clubs or could be found at universities. Several years passed before we had graphics capability. I started computing from home by connecting through the phone lines to the university computer center with a dumb terminal. My wife was taking a course in computing, and we had to drive to the computer center to pick up print outs. Our modem was so slow that there was hesitation as each character was typed. A dot matrix printer was soon connected to the spare port on our dumb terminal, and not so many trips to the computer center were needed. Another computer purchased for home used our dumb terminal for display and led to mostly local computing, with the university center available when needed. As faster modems became available, we upgraded for better service. By about 1982, I was using electronic communication to colleagues at other institutions. Software was becoming available for entertainment that provided breaks from serious programming. My wife became a publisher because my books were integrated with teaching programs on a disk, and major publishers were leery about distributing disks and providing customer support for the programs. The university now had a laser printer that we used to make camera-ready copy for my books. My wife learned to use some packages for preparing manuscripts and eventually found that LaTeX was wonderful. The LaTeX commands for spacing terms in an equation are complicated, and I remember how she spent hours getting one messy equation to print correctly.

The Apple computer with full color display when connected to a television set showed what a personal computer could be. Its popularity encouraged competition that brought the price of crude home computers to as low as $100. Some people in the sciences and in engineering used the Apple computer professionally, but it was not quite right. It was clumsy for editing text because letters large enough to read on a TV screen required truncating the line to only 40 characters. You were better off connecting your computer to a monitor with good, readable, full lines of text. The early IBM computers and the many clones that were soon available had only a monochrome display, but the monitors were easy to read.

BASIC can do just about anything and is nicely suited to personal computers. It has ways to get signals from a port and to send signals back. Early FORTRAN for personal computers did not come with easy ways for reading and writing to the ports. When most programs were small, it did not matter so much that BASIC was slow. Its interpretative code runs right away, and FORTRAN and the other powerful languages require a compiling step.

Interaction with the computer was at the command line at which you typed your instruction. The graphical user interface was popularized by Apple Computers and was a sensation with the monochrome Macintosh. While the Apple company kept close control of its system, IBM used the DOS operating system that made Bill Gates a billionaire. This was an open system that led to many companies competing to provide software. Apple has done well in some niches for software, but PCs that developed from the IBM system have a richer array of software that has driven them to a predominant share of the market.

I went a different route in the early 1980s with the Commodore Amiga, a truly magnificent machine that was badly marketed. The Amiga was fast and great for color graphics because it had specialized chips to assist the central processor. It had both a command line interface and icons. At one time, I had five Amiga computers at home, in my office, and in the laboratory. I used the command line perhaps a little more often than I clicked on an icon. With today's Windows, it is not worth the trouble of opening a DOS window so that you can use a command line and wildcards to make file transfers easy. The Amiga had true multitasking. This required about 250 kilobytes of memory in contrast to today's multitasking systems that gobble memory and require about 80 megabytes of your hard drive. My first Amiga crashed a lot, but later models did not. My computer purchased in 1998 has the Windows operating system and crashes two or three times each week.

Minicomputers evolved into workstations and developed side-by-side with personal computers. Magnetic storage started with large drums or disks and became smaller in size, larger in capacity, and lower in price. Persistent memory chips stored programs to get the computer up and running. Eight-inch floppy disks were rendered obsolete by 5-1/4-inch floppies that gave way to the 3-1/2-inch disks. The first PCs with hard drives had only 10 megabytes. My first Amiga with a hard drive (70 megabytes) made dismaying noises as it booted. Inexpensive personal computers now have options of multi-gigabyte hard drives. I find essential a Zip drive with 100 megabytes of removable storage. There are devices with much more removable storage, but I find it easier to keep track of files when the disk does not hold too many.

It was a logical step to use the ability of the computer as the basis for word processing. With the early programs, you could only insert and delete on a line-by-line basis. The next advance was imbedded commands that controlled the printed page. I was served very well for about seven years by TeX and its offshoot LaTeX that had a preview program to show what your pages would look like. What-you-see-is-what-you-get seems so unremarkable now, but it revolutionized word processing. The version of LaTeX for the Amiga came with over a dozen disks with fonts, but there were very few types. These were bit-mapped fonts, and each size and each style required a different file on the disk. I obtained fonts at computer shows, bought some Adobe fonts, and found others in archives at universities. These were intended for PCs, but the files were recognized by my Amiga computer. I had to install them on my hard drive and learned how to send them to the printer. Proportional fonts that are scaled by equations have made my a huge collection of bit-mapped fonts obsolete. There was also incompatibility between PostScript and other printers, but conversion programs solved this problem.

It may seem extraneous to focus so much on the hardware and software, but your use of a tool depends on its capabilities. New users today cut their teeth on word processing, perhaps as part of their e-mail, but this was NOT a common use of computers in the early days. There were few CRT displays except at the computer center itself, and users worked with printed pages of output that were often just long listings of the programs for debugging. These were big pages, and printing on letter-size paper seems not to have occurred to anyone.

Many of us realized that pictures are better than words and wrote programs that showed our students not only columns of numbers but also pages with Xs, Os, and other characters positioned as a graph on the print out. Better graphs were available from a line printer, but there were few of these, and it was troublesome to walk some distance to get your results. There was usually a charge associated with using a line printer, and someone had to make sure that its pens had ink and were working. There is a great difference between computer output as printed lines of alphanumeric characters and output as drawings and graphs. It was quite some time before the affordable small printers for personal computers had graphics capability, but monitors for graphics became common. Furthermore, the modern computer can update and animate its images for its CRT display. BASIC for our computers had powerful graphics calls that were easy to learn. The professional programmers used languages such as C for high-speed graphics. Programs for word processing were followed by spreadsheets and other business programs. With the advent of games, the software industry took off.

3 Biotechnology

Portions of this historical review pertain to academic computing in general, but there were some specific features for biotechnology. Three interrelated areas of particular importance are simulation, process monitoring, and process analysis.

3.1 Simulation

Simulation, an important tool for biotechnology, is considered essential by many bioprocess engineers for designing good control [5]. As you gain understanding of a system, you can express relationships as equations. If the solution of the equations agrees well with information from the real system, you have some confirmation (but not proof) that your understanding has value. Poor agreement means that there are gaps in your knowledge. Formulating equations and constructing a model force you to view your system in new ways that stimulate new ideas.

Modeling of bioprocesses had explosive growth because the interaction of biology and mathematics excited biochemical engineers. Models addressed mass transfer, growth and biochemistry, physical chemical equilibria, and various combinations of each of these.

It becomes impossible to write simple equations when an accumulation of factors affects time behavior, but we can develop differential equations with terms for important factors. These equations can be solved simultaneously by numerical techniques to model behavior in time. In other words, we can reduce a system to its components and formulate mass balances and rate equations that integrate to overall behavior.

The concept of a limiting nutrient is essential to understanding biological processes. The nutrient in short supply relative to the others will be exhausted first and will thus limit cellular growth. The other ingredients may play various roles such as exhibiting toxicity or promoting cellular activities, but there will not be an acute shortage to restrict growth as in the case of the limiting nutrient becoming exhausted.

The Monod equation deserves special comment. It is but one proposal for relating specific growth rate coefficient to concentration of growth-limiting nutrient, but the other proposals seldom see the light of day. This equation is:

$$\mu = \frac{\hat{\mu} S}{Ks + S} \tag{1}$$

where

μ = specific growth rate coefficient, time^{-1},
$\hat{\mu}$ = maximum specific growth rate, time^{-1},
S = concentration of limiting nutrient, mass/volume, and
Ks = half-saturation coefficient, mass/volume.

Students in biochemical engineering tend to revere the Monod equation, but practicing engineers apply it with difficulty. There is no time-dependency; it is not a dynamic relationship and cannot handle sudden changes. Industrial batch processes encounter variations in the characteristics of the organisms during the run such that coefficients on the Monod equation must be readjusted.

Simulation paid off. One of my students, Thomas Young, joined Squibb in about 1970 and soon made major improvements in the yields of two different antibiotic production batches, mostly as the result of simulation. I had recommended Tom to my old employer. They declined to make him an offer because they considered him too much of a theoretical type. A vice-president at Squibb told me that Tom was just about the best person that they ever hired and that his development research saved their company many millions of dollars. It was partly the ability to test ideas on the computer that led to rapid progress, but even more important was the thought process. Deriving equations for simulation forces you to think deeply and analytically, and many new insights arise.

3.2 Monitoring and Control of Bioprocesses

Instrumentation in a chemical plant brings to mind the control room of a petroleum refinery with its walls lined with a cartoon representation of the processes with dials, charts, controllers, and displays imbedded at the appropriate locations. Operators in the control room observe the data and can adjust flow rates and conditions. Such control rooms are very expensive but are still popular.

There are hybrids of the traditional control room with computers for monitoring and control. One computer monitor has too little area for displaying all of a complicated process, and the big boards along the walls may still be worthwhile. However, smaller processes can be handled nicely on one screen. Furthermore, a display can go from one process unit to another. A good example is the monitoring of a deck that has many bioreactors. One display can be directed to a selected reactor to show its status, often with graphs showing variables vs time. The operator can choose the reactor of interest, or each one can come up for display in sequence. Logic can assign high priority to a reactor that demands attention so that it is displayed while an alarm sounds or a light flashes. If we agree that an operator can only watch a relatively small amount of information anyway, it makes sense to conserve financial resources by omitting the panel boards and using several computer displays instead. There is the further advantage that the computer information can be viewed at several locations; some can be miles away from the factory. Connecting from home can save a trip to the plant by a technical person at night or at the weekend.

I think that bioprocess companies have done well in using computers to monitor their processes. The control rooms are full of computers, but the adjustments tend to be out in the plant. Only recently have plant operations been fully automated or converted to control by sending signals from the computers. Because hardware and the labor to set it up and wire the connections have a daunting cost, the lifetime of a computerized system is roughly 5 years. Advances in the technology are so rapid that a system after 5 years is so obsolete that it is easy to convince management to replace it.

I have an anecdotal report of early attempts at automation. The Commercial Solvents Corporation in the 1950s attempted to automate sterilization of a reactor. There was no computer, and the method relied on timers and relays. Unreliability due to sticky valves and faulty switching resulted in failure. Being too far ahead of their times gave automation a bad name. Development engineers at other companies who proposed automation were told that it had been tried and did not work. Today, there is nothing remarkable about computerized bioreactors and protocols for their operation. I have observed automated systems in industry that are fully satisfactory. The computer makes up the medium and can control either batch or continuous sterilization.

3.3 Bioprocess Analysis and Design

There is not a great difference now from what was being done in the past, but there have been many changes in convenience and in capabilities. Engineering professors wrote programs that assisted process design with such features as approximating physical properties from thermodynamic equations. These properties are crucial to such tasks as designing distillation columns but do not matter much in biochemical engineering. Today, there are excellent commercial programs such as Aspen that will develop the required thermodynamic properties en route to designing a process step or even an entire system. I experimented with Aspen and my opinion of it comes later.

4 Recent Activities

A major advance has been databases and programs that manage databases. Libraries of genetic sequences have become essential to many biotechnologists, but this area deserves its own review and will not be mentioned again here.

4.1 Models

Models should be judged on how well they meet some objective. A model that fails to match a real system can be highly valuable by provoking original ideas and new departures. Overly complicated computer models, of course, can have a fatal weakness in the estimation of coefficients. Coefficients measured in simple systems are seldom valid for complex systems. Often, most of the coefficients in a model represent educated guesses and may be way off. Complicated models take years to develop and may be impractical to verify. Such models are worth something because of the organized approach to just about all aspects of some real system, but there are so many uncertainties and so many opportunities to overlook significant interactions that predictions based on the models may be entirely wrong.

Deterministic models (those based on actual mechanisms) make a great deal of sense when they are not too unwieldy. The terms have physical or biological meaning and thinking about them may lead to excellent research. The goal of the modeler should be to identify the most important effects and to eliminate as many as possible of the insignificant terms. It always comes back to the purpose of modeling. To organize information, we may just keep adding terms to a model in order to have everything in one place. When the goal is prediction, a model should be tractable and reliable. That usually means that it must be simple enough that its coefficients can be estimated and that the model can be verified by comparing its predictions with known data.

Most real-world situations are too complex for straightforward deterministic models. Fortunately, there are methods that empirically fit data with mathematical functions that represent our systems and permit comparisons and predictions.

Models that assume steady state are usually fully satisfactory for continuous processes. Continuous cultivation tends to steady state or may fluctuate fairly slowly so that the Monod equation can be applied. Batch processes and dynamic situations in continuous processes must consider non-steady states. Franzen et al. [6] developed a control system for respiratory quotient and found that steady-state models were unsatisfactory.

The term *hybrid* brings to mind hybrid computers that combined an analog computer and a digital computer. The term *hybrid model* is applied when a model has components that attack a problem quite differently. For example, there might be deterministic features, aspects of neural networks, and optimization routines that must be organized and integrated [7, 8].

Computer models aid state estimation for bioprocesses. Such models have great value for predicting the best harvest time for a bioreactor when the product concentration peaks and declines or when the rate of product formation no longer repays the cost of continuing the run.

4.1.1 *Unstructured Models*

The simplest models lump relationships. A good example is letting X stand for cell mass. This is an absurd oversimplification because cell mass is composed of hundreds of biochemicals that can change in proportions and that function very differently. Nevertheless, growth dynamics may be represented fairly well by the simple equation:

$$\frac{\mathrm{d}X}{\mathrm{d}t} = \mu X \tag{2}$$

Letting S stand for the concentration of growth-limiting nutrient is another example of lumping because a real nutrient medium can have several ingredients that contribute. For example, the intermediary metabolism of fats produces compounds for carbon-energy, but we may focus on a sugar as the limiting nutrient.

4.1.2 *Structured Models*

Ramkrishna's group at Purdue University uses the term *cybernetic* for biological models that interpret the control actions of organisms [9, 10]. Current interest seems to be focused on simultaneous or preferential uptake of carbon sources. By incorporating cybernetic concepts, they can predict diauxic growth patterns as well as simultaneous use. The structured model considers precursor compounds and the patterns for their formation and consumption.

Structured models can become complex when the fine details of growth, biochemistry, and equilibria are incorporated. One example is a description of *Escherichia coli* developed under the leadership of Shuler at Cornell University [11] that has dozens of biological effects, a multitude of chemical reactions, and hundreds of coefficients.

A fungal bioprocess modeled in great detail qualifies as a structured model even though some aspects of the biology are lumped [12]. For example, fungal morphology (such as pellet size and the fraction of pellets in the total biomass) is a structural feature that could be further categorized for the biochemical details.

4.2 Bioprocess Control and Automation

Appendix 1 has some terminology of process dynamics and control for readers who are not engineers. The theory of process control stays ahead of practice, and some aspects of the theory have minor importance for biotechnology because our processes change slowly and have no analogy to exothermic chemical processes that can explode. There are still times when it makes sense to purchase a conventional controller with proportional-integral-derivative control. For example, a storage tank may benefit from temperature control, and an inexpensive controller that adjusts the flow of cooling or heating water through a jacket or coil may suffice. However, most factories have computers that control their bioprocesses, and there are enough places to hook up sensors and to send signals to controllers to accommodate some more control. When you run out of channels, you buy another interface board or another computer. The cost can be far less than purchasing several controllers, and the entire control system can be displayed conveniently in one location instead of at the individual controllers that may be widely scattered. Sonnleitner [13] provided perspective on automation and design of bioprocesses and addressed some potential pitfalls. Galvanauskas et al. [14, 15] have a simple procedure for enhancing bioprocess performance through model-based design.

The concepts of proportional, integral, and derivative control retain their importance, but using a computer in place of a controller permits far more sophistication [16]. The settings for the gain of the controller and its coefficients for its various actions are compromises because the process can change. This is particularly true of bioprocesses. Batch processes start with high nutrient concentrations, few organisms, and distinct properties such as temperature, pH, viscosity, and the like. As the run progresses, concentrations change as do physical properties. The controller settings that worked well early in a run may be poor at a later time. The control algorithm in the computer will employ concepts of proportional, integral, and derivative control, but can augment them with logic. In the simplest case, past experience may be used to devise a schedule for changing the control coefficients in steps, e.g., when the time reaches some point, and switch to these coefficients. In the more advanced cases, a model of the process makes the decisions for instantaneous adjustments of the control coefficients.

4.2.1
Sensors

Manual sampling at frequent intervals was the norm for industrial bioprocesses for many years. Gradually, more and more methods have been developed for on-line measurements. Until the early 1960s there were no reliable pH electrodes that could be sterilized. Several of our bioprocesses performed best at the correct pH, and we put a great deal of effort into taking samples, titrating, and adjusting the pH in the vessel. The first sterilizable electrodes had short lives. If the pilot plant had 20 vessels with internal pH electrodes, half might fail during a run. Although expensive, an internal pH electrode is standard operating procedure today.

The extensive research and development of better sensors for bioprocesses is beyond the scope of this review [17]. One example of the level of sophistication of current procedures is Rothen et al. [18].

4.2.2
Observers

An observer is a method for mixing models and real data. The general concept is that some variables can be measured by cost-effective analytical procedures while other variables are costly, troublesome, or impractical to measure. The observer uses both the model and the practical measurements to estimate those variables that were not measured directly. An example of something that is impractical to measure continuously is cell mass. The usual estimate of cell mass draws samples, collects the solids by filtration or centrifugation, and dries and weighs them. This is time consuming and laborious, so the value for cell mass is unavailable when needed as a control variable. There are alternative methods of estimating cell mass by optical density, nucleic acid content, or the like, but such methods track imperfectly with cell mass. A good model of the bioprocess will have cell mass as one of its terms. We can construct the model so that it uses the measured variables and estimates cell mass. If the model is reasonably good, we can use the estimate of cell mass as an index to control the process.

Bioprocess variables that are usually measured continuously are pH, temperature, and feed rates. Other possibilities are impeller speed, aeration rate, concentrations of carbon dioxide and oxygen in the exiting gas, and perhaps some concentrations in the medium. Some textbooks refer to the assays that are used for estimation of other parameters as *gateway* sensors because they open the door for using the model.

Models are not perfect representations of a process. As the bioprocess and its model drift apart, estimates made by the model will be incorrect. Errors propagate with time and become more serious. However, we can measure critical variables occasionally and correct the estimates. An analogy would be an internal pH electrode that drifts haphazardly. Before the next run, we would replace the failing electrode with a good electrode, but the bad electrode may be all that we have right now. We can still use its readings if we take samples from the bioprocess, measure the pH accurately at the lab bench, and adjust the reading of the poor electrode. If the drift is very large, we must recalibrate more often. In any event, the pH signal gives us something for continuous control of the process, and control based on grab samples is a very unattractive alternative. When resetting the pH reading, we should correct to the time that the sample was taken, not to the present time because the electrode may have drifted during the sampling.

The above method is also applied to observers, but there are other ways to correct. One useful technique is to estimate something that can be measured continuously. The estimates are compared with the true values to decide how much the model is in error, and the amount of error determines the compensation for other estimates. This is really just a slight difference from using the additional measurement as part of the model.

Off-line measurements are of limited value for control but can be major assets for checking a model and correcting estimates by an observer. Poor models are risky. If the model is unstable, model errors can give rise to estimation errors that increase exponentially. This is an example of when bad control is worse than no control.

4.2.3 *Auxostats*

The concept of an auxostat is based on process control. The nutrient feed stream is used to fix some variable of the process. Pumping dilutes concentrations in the medium. If the microbial population is unable to maintain the desired concentration, the pump is automatically slowed down while the organisms both grow and excrete product. Once steady state is established, the specific growth rate of the organisms equals the system dilution rate [19]. I think that Edwards et al. [20] were the first to publish a description of an auxostat, but Watson [21] previously did something quite similar. The turbidostat, first used in the 1940s, has this same concept, but the property being controlled is turbidity of the culture.

Auxostat operation of a continuous reactor system holds the promise that the culture can be operated safely at a high specific growth rate [22]. When, for example, the product is cell mass, dilution by fresh medium is offset by growth. Furthermore, there is intense competition among the organisms. This operating mode has the advantage of washing out the slower growing microorganisms from the reactor leaving more substrate for those that grow faster. It is also obvious that a faster growing population (from mutation or contamination) challenges the established culture to cause its loss from the system. Auto selection of yet faster growing cultures is the logical result. When the specific growth rate should be fixed and not allowed to seek its own maximum, the control method of Levisauskas et al. [23] can be used.

4.2.4 *Examples*

Some typical examples of computerized control are shown in Table 1.

Table 1. Examples of computer control

Product or Process	Strategy	Reference
Glutathione	Advanced control	24
Acetate	Internal model	25
L-Carnitine	Analysis of model	26
6-Hydroxynicotinic acid	Analysis of model	26
5-Methyl-2-pyrazincarbonic acid	Analysis of model	26
Nicotinamide	Analysis of model	26
Alkaline protease	Better monitoring	27
Acetone/butanol	Metabolic analysis	28
Penicillin	Hybrid model	29
Penicillin enzymatic deacylation	Kinetics, equilibria	30
Lipase	Growth rate control	31
Brewers yeast flocculation	Analysis of model	32
Waste water treatment	Neural network	33

4.2.5
Modeling and Control of Downstream Processing

Some aspects of recovery of products after bioprocessing are relatively straightforward. Admittedly, there are complications due to chemical equilibria and reaction kinetics, but these are not unique to processes for purification of product streams. Downstream processing incorporates the main lines of chemical engineering and can draw on a wealth of experience. Software tools and models were discussed by Simutis et al. [34]. The commercial programs, such as Aspen, that are used to design chemical processes can be used as well for downstream processing, but I have not been impressed very much by their applications to biochemical engineering. One beauty of such programs is handling of physical properties. Essential coefficients can be drawn from a library that is part of the program, or they may be estimated with thermodynamic equations. While this is a great aid for the usual chemical process industries for such steps as distillation, there are not many biochemical engineering steps at high temperatures or pressures where you need to know the thermal properties of the process streams. Practically all of biotechnology is aqueous.

Having stated that there is little difference from regular chemical engineering in terms of the intellectual challenges of computer approaches to downstream processing, I must admit that in almost all cases, the recovery operations cost much more than the bioprocessing. It makes a great deal of sense to improve costs through better modeling and control of downstream processing. Bulmer et al. [35] have applied computer simulation for improving the recovery of intracellular enzymes and demonstrated this with alcohol dehydrogenase from *Saccharomyces cerevisiae*. Zhou et al. [36] modeled the interrelationships of bioprocessing and product recovery to optimize both. Vanhalsema et al. [37] modeled phase equilibria to improve extraction of carboxylic acids.

4.3 Intelligent Systems

State-of-the-art control systems employ artificial intelligence. Several companies now offer complete systems for monitoring and control of bioprocess, and the control algorithms use advanced logic. Two web sites for further information are http://www.biospectra.ch and http://www.gensym.com. One of my former students worked for GenSym and played a major role in developing such a system. The following descriptions of current practice are based on pages downloaded from their web site.

Example 1: Gensym's G2 Diagnostic Assistant (GDA) was installed in the fermentation plant at Eli Lilly's Tippecanoe Laboratories (Lafayette, IN) with three goals: control its antibiotic production processes by reducing variability, increase yield while reducing unit costs, and give operators the responsibility for the process. Using statistical information from previous runs, control charts were developed. This system monitors incoming production data, compares them against the established control charts, and provides operators with expert advice. A distributed control system collects and processes data, provides expert instructions in real time, and also provides an intelligent operator interface that advises the operator on how to respond to unusual behavior or informs the operator that the situation is sufficiently novel that no advice can be offered.

In the first month of system operation, process deviations dropped 72.3% and yields increased 3.2%. As a result, the company has realized significant cost savings and has become better able to match production to changes in sales.

Example 2: The Bioindustrial Group at Novo Nordisk (Denmark) uses a Gensym system for monitoring, sensor validation, diagnostics, and state estimation of fermentation processes for the laboratory scale production of industrial enzymes produced by genetically engineered microorganisms. Suitable sensors may not be available to determine the metabolic state of the processes, so the system combines sensor data with analytical data for state estimation of the culture's metabolic status. Bioprocesses have many problems with sensors. This system continuously validates sensors using statistical calculations and mass balances and identifies deviations from normal or expected behavior while providing high level supervisory control. Controller "responsiveness" is an important aspect of Novo applications, and the system makes the control more responsive to variations in the culture using empirical on-line optimization.

Stephanopoulos and Han [38] have reviewed intelligent systems in process engineering. Quite involved or complicated logic can be programmed. Instead of fixed setpoints or controller settings, there can be a schedule for adjustments, or logic can decide the settings. Furthermore, artificial intelligence can analyze the process to establish the control specifications. Diaz et al. [39, 40] have used predictive control of the concentration of dissolved oxygen in a bioreactor.

4.3.1
Expert Systems

Control can be based on rules in an expert system. For example, an expert system can take into account the lag phase and the youthful physiology of the cells for control of the early stage of a bioprocess while the control can be much different for other phases. Programming simple rules as IF/THEN statements is not a serious challenge for control engineers, but commercial programs called *Expert System Shells* can usually do it better. These shells provide for compound rules, rule priorities, exceptions, explanations of why certain rules have fired, and more. One important report is the rule activity chart that tells which rules fired, when they fired, and how intense was their action. This helps greatly when revising and improving rules. Furthermore, comparing rule activity of one run with another run can improve understanding of the process and the ways for improvement.

De Bernardez et al. [41] developed an early expert system for controlling a bioprocess. It was particularly interesting that the expert rules had to be selected carefully to avoid exhausting the available memory of the computer. This seems archaic now when personal computers have enormous sizes of memory, but not long ago memory could seriously limit what could be done. The maxim "work expands to fill the available time" has the analogy that computation can expand to tax the available memory. For example, collecting data from several bioprocesses at frequent intervals can create gigantic files. Say that the sampling rate is 100 times per second and that a run lasts for 100 h. If there were 10 sensors to be sampled and 10 calculated points to be plotted, the file size to hold all this information would be:

$$100\ \text{h} \times 60\ \text{min/h} \times 60\ \text{s/m} \times 100\ \text{times/s} \times 10\ \text{sensors} \times 10\ \text{calculations} = 3.6 \times 10^9$$

It is not wise to use memory resources this way; the data should be averaged for a time interval. For example, the average pH during 1 min of the run may be adequate with no need for knowing its value each one hundredth of a second. If the file is still too large, it should be compressed.

4.3.2
Fuzzy Logic

One organized method for dealing with imprecise data is called *fuzzy logic*. Even with precise information, real-world problems may include uncertainty. There is a range of values for failure, and the designer of a bioreactor will incorporate safety factors to make collapse or rupture highly unlikely. Unreliable data magnify the chances for erroneous conclusions, and fuzzy algorithms commonly estimate the level of confidence of the results. The data are considered as *fuzzy sets*. Traditional sets include or do not include an individual element; there is no other case than true or false. Fuzzy sets allow partial membership. These memberships can be manipulated with the operators shown in Table 2.

As would be expected, probability is often needed to clarify fuzzy, uncertain information. A degree of truth is the probability that a statement is true. For a

Table 2. Fuzzy operations

Operation	Definition
Fuzzy AND (f-AND)	a f-AND b = min(a, b)
Fuzzy OR (f-OR)	a f-OR b = max(a, b)
Probability AND (p-AND)	a *p*-AND b = a*b
Probability OR (p-OR)	a *p*-OR b = ab – (a*b)
(same as Unary NOT)	NOT a = 1 – a

practical situation, we might use one of the following methods to design a fuzzy controller [42]:

1. Base the controller on a human operator's experience and/or a control engineer's knowledge
2. Model the control actions of a human operator
3. Base control on a fuzzy model of the process

The first method is analogous to constructing an expert system, and this is the most common approach to fuzzy control. The first and second methods can only approach the performance of a human because they strive to mimic a human. Some processes are so complicated that no humans can control them well. For example, the U.S. Air Force has a fighter plane that cannot be flown by hand – the computer must fly it with the human supplying information about where to go and how fast. To go beyond human capability, Method 3 with a fuzzy model of the process has the best chances for success. The problems are to identify the system features in a form that may be modeled and to design an effective controller that can be applied to this model.

Fuzzy reasoning can incorporate heuristic knowledge into the model. Simutis et al. [43] report using fuzzy information with an extended Kalman filter to follow alcohol formation in beer brewing. Their model outperformed conventional approaches for estimating the process state of an industrial scale reactor. Their linguistic variables had the following forms:

- Process time: short, medium, long
- Sugar concentration: low, medium, high
- Process temperature: low, medium, high
- Sum of evolved CO_2: small, medium, high
- CO_2 evolution rate: low, medium, high

Truth tables were developed to assign membership in these categories. Precision of some of the assay methods was low, and the fuzzy decisions were more appropriate than using the numerical values. More recent advances incorporate state modeling with fuzzy control [44].

4.3.3
Neural Networks

Neural networks represent knowledge in an adaptive, distributed architecture [45]. The general idea is relating outputs and inputs in a manner akin to *pattern recognition*. There are input nodes and output nodes with one or more layers of nodes in between. Weighting coefficients for the summation of inputs to each node develop the outputs. The inputs and outputs for sample cases are used to improve coefficients through iteration, and eventually the system can derive reasonably correct outputs from inputs never before encountered.

A crude example of how a neural network could be useful is a process with several different feed streams and measurements such as concentration, temperature, and pH. There will be permutations of our input information for which the output is known. Output may be action rules such as add more acid, increase the rate of cooling water, sound an alarm, or the like. As our neural network iterates and learns with solid examples of appropriate outputs for given sets of inputs, it can evolve until it recognizes patterns in these inputs. When presented with totally new inputs, it may be able to generate a correct output. The correctness of the output will depend on how well it has been trained, but insufficient examples from which to learn will give unreliable answers.

Important facts about neural network programs are that learning is slow (medium-sized networks may take hours of training on a slow computer) but decisions with a trained network can be lightning fast. The learning requires iteration, error checking, and testing for convergence. Calculating the output of a trained network is merely once through with multiplication and addition. Commercial programs called *shells* are easy to use with no mathematics. Furthermore, the neural network learns automatically; you configure it but do not have to write any program. The neural network tends to be as good as its training data. A well trained and carefully tested network is a highly reliable system that did not need many long hours of programming by a human.

Vancan et al. [46, 47] have presented strategy for using neural networks in bioengineering situations. Simutis and Lübbert [48] discussed developing process analysis through mass balances and neural networks and augmented such analysis with random search optimization routines that find optimal regimes of feed rate, feed concentration, temperature, or precursor profiles [49].

Stoner et al. [50] have used computer interfacing and control with integrated software packages including learning to deal with multiparametric effects on the oxidation of iron by thermophilic bacteria. Shimizu [51] has combined fuzzy logic with a neural network and provided three examples of processing with *Saccharomyces cerevisiae* that used computerized state recognition and fuzzy control of the processes.

4.4 Responses of Microbial Process

Steady state analysis is becoming outdated because there is no steady state in a batch process while the organisms are alive, and control of continuous processes must consider dynamics to be effective. A report that it took time for microorganisms to respond to forcing by Mateles et al. [52] affected my thinking very nearly at the start of my professional career, but a decade later there were still grievous errors in dynamic analysis. The same Thomas Young that I lauded earlier presented a paper at a national meeting of the American Chemical

Society in about 1970 and upset the audience by criticizing the Monod equation. It seems so obvious now that this important equation has no time dependency. The specific growth rate cannot respond instantaneously, so the equation is not equipped to handle abrupt changes in nutrient concentration. Nevertheless, it was heresy at that time to criticize one of the tenets of biochemical engineering.

With no delay in growth-rate adjustment, the response of a bioprocess to an upset should move smoothly to the new final value without overshoot. Real bioprocesses tend to overshoot because organisms require time to adjust their biochemistry to new environmental factors.

Most university computer centers have programs that aid in dynamic analysis, and some programs for PCs have appeared. The user can specify the scheme of a system (this is termed the *block diagram*) and the time constants. Either the connections of blocks or the coefficients can be changed to determine the response and stability. There are also features for analyzing stability.

4.5 Metabolic Engineering

Metabolism and biosynthesis depend on parallel and sequential biochemical reactions subject to controls based on induction, activation, or inhibition of the required enzymes. Intermediate biochemicals progress through competing pathways at rates determined by the activities of these enzymes. The manipulation of these enzyme activities is termed *metabolic engineering*. Some control is possible by adding precursor compounds or substances that affect enzyme synthesis. For example, a number of carbohydrates induce or stimulate the formation of cellulase enzymes. Adding them will result in higher activity of cellulases. However, most metabolic engineering comes from genetic alterations to cripple certain unwanted pathways or to promote desired pathways. Computer models of the reactions and biochemical controls aid research by highlighting the enzyme activities of interest and predicting the effects of changes.

Metabolic engineering is one of the most exciting areas of biotechnological research but is beyond the scope of this review. A few typical references can be consulted [53–57].

5 Information Management

My years in industrial process development were in the dark ages of data processing. We collected hand-written records kept by the operators of the results of samples taken for pH, turbidity, packed cell volume in a centrifuge tube, temperature, amount of antifoam added, concentration of product, and the like. We requested assays for sugar concentration, amino acids, and specific biochemicals when appropriate. Everything was on paper and circular charts or strip charts. To compare one run with another, you had your assistant and maybe other people covering one or two tables with paper, sifting through, and calling out information from one run as you inspected another. This was such a chore that we did a rather poor job of tracking the history of runs to learn how the best runs differed from the others.

With a computer interfaced to a bioprocess, the collected data are already in a format suited for a spreadsheet or database program. These programs have features for displaying results as graphs or charts. Commercial database programs make it easy to compare different experiments and runs. These programs include routines that manipulate and treat the data. We can easily set graphs side-by-side or even superimpose them. There is no excuse for not searching for reasons why some runs are better than others. Developing better ways to search these databases is an active area of research [58].

Networking of computers ranges from a few stations to the whole world. Laboratory information management systems (LIMS) substitute computer actions for some routine, boring human labor and eliminate errors from entering and transcribing data.

5.1 Customer Service

Some companies have gone far beyond routine advice to customers by phone, fax, e-mail, or personal visits by integrating their computers with those of their customers. A good example is Iogen Limited, based in Ottawa, Canada. They get on-line reports from computers that control customers' reactors that use Iogen enzymes. When there are process deviations, recommendations are made almost immediately. There was an anecdotal report of another company attempting to gain market share from Iogen by offering similar enzymes at one-half the price. The customer declined that offer with contempt, and the manager who had hoped that price alone was the key to sales was fired.

5.2 Electronic Communication and Teaching with Computers

All areas of science and engineering make use of e-mail, fax machines, and the Internet. Information today flows at much faster rates than in the past and in huge amounts. Graphic information benefits from computers. Digital cameras and scanners are inexpensive. We can include color images in our communications very easily. Color printing that was costly just a few years ago is now routine with printers sold in computer stores for as little as $150.

Biochemical engineering is well-served by web sites that make our lives easier. In preparing this review, it took but a few hours to search for the authors whose research I had been following and for keywords to compile examples of current work. Our computer programs allow cutting and pasting from one to another. I collected references as files from the Internet and cut out the needed information with little retyping.

I strongly support electronic publication, but the emphasis so far is on better distribution. Short times between submission and publication are important advantages of handling manuscripts with e-mail, and placing reports that look much like the old hard copies on the Internet insures wide and timely distribution. However, electronic journals in their infancy have hardly begun to use modern technology. I think that an exciting development is pages or displays that do something, and this was my main reason for learning Java.

Science and engineering are sprinkled with equations. Graphs based on equations are appreciated and convey information more effectively than just presenting the equations and their terms and symbols. Books and articles cannot afford very many figures, but computer code has no limitations on the number of graphs it draws. By programming to draw graphs of an equation based on reader input, students can explore permutations of its coefficients, learn quickly, and retain a feel for which terms matter most and how sensitive is the result to each coefficient. My educational projects have led to web pages that can be accessed from all over the world. Having my Java code as a template, I can create an interactive web page with a different equation in 20 min. I have extended this to simulations of microbial processes for which the reader can manipulate the parameters.

Examples are applets with scrollbars that change the coefficients for graphing of normal distributions, the Monod equation, adsorption with either the Langmuir equation or the Freundlich equation, Michaelis-Menten enzymatic kinetics, and various growth rate expressions for multiple nutrient limitations. Several of my applets mimic an experiment. The reader selects some conditions, and the computer displays the expected results. Other portions of the page can suggest things to try and what to observe. There can be links to explanatory material. Some of my applets show solutions of simultaneous ordinary differential equations such as dynamic responses to various forcing functions and interactions of microorganisms.

Games grab student interest. By keeping alert to opportunities for teaching with a game, I have had about 5 successes in 38 years. You have to decide what you are trying to teach before you devise some complicated game. An approach that has worked for me is to develop some sort of role playing and to award hypothetical profits that depend on how well you understand your role. Students appreciate money and enjoy making even simulated profits.

Electronic documents can have all sorts of motion, even video if you are willing to wait while it downloads. Sound with the video is seldom important for biotechnology, but a voice could add a personal touch. You do not need Java to present animations. A number of methods depend on shuffling through a collection of images to make a movie. Java also has features for doing this. When simple shapes will suffice for figures, Java subroutines can easily move them around the screen for what I call *algorithmic animation*. One of my animations that required 1 megabyte for separate images was replaced by about 5 kilobytes of Java code that sketched the same moving images (a 200-fold saving). Since this code runs fast right at your local computer, you wait only a moment to download the small Java applet before the animation starts. My Java applets can be found at http://www.rpi.edu/dept/chem-eng/biotech-environ/biojava.htm

6 Some Personal Tips

Modern bioprocesses are seldom operated in strictly batch mode where nothing but air is added after inoculation. Certainly for research there will be some additions of nutrients, reagents, or special biochemicals. Such additions require pumps, and small, reliable pumps are not inexpensive. An adjustable pumping rate is more desirable than a fixed rate, but variable rates pose problems for the computer. Some very nice pumps with variable speed motors are commercially available. These require special control boards that plug into the computer or the interface may connect to one of the ports of the computer. The computer sends one signal to the pump to set its speed and another signal to turn the pump on or off. We have had excellent experiences with these adjustable pumps except that their cost is relatively high.

Constant speed pumps are less expensive than variable pumps, but the rate may be matched poorly to the need. A brief interval of pumping may overshoot the set point and upset the process. We have solved this by establishing a pumping cycle of the order of 10–20 s, an interval that appears nearly continuous on the time scale of microbial processes. Our computer turns a constant speed pump on for a fraction of this interval. For example, to pump at half the maximum rate of the pump, it would be turned on for 50% of the cycle. The electronics for doing this cost very little, and the parallel port of a computer may have sufficient power for the switching. If not, one transistor can develop the needed power. An optical relay on the computer side of the circuit isolates the pumping circuit. When the computer turns on a tiny light-emitting diode in the relay, voltage is applied to the pump. These relays cost from $1 to $10 depending on how much current must flow. Heating circuits can use this same principle and at far lower cost than analog controllers of high wattage that vary the voltage to a heater.

Stepping motors are another way to get variable speed. Each pulse sent to the motor control circuit turns the motor a few degrees. The motor speed can be geared down to get a small fraction of a degree per pulse. This is a superb method for getting variable speed for low torque situations, but heavy-duty stepping motors are expensive. Stepping motors for light duty are ubiquitous for moving paper through printers and are a quite inexpensive. We have enjoyed the excellent performance of stepping motors connected to personal computers to drive home-built micromanipulators for our research with microelectrodes inserted specific distances into microbial films. We used an O-ring as the belt to link the stepping motor to the fine adjustment of a microscope body to which our microprobe was fastened. Each pulse from the computer raised or lowered the microprobe by a fraction of a micrometer. The programming to send pulses to the parallel port of a personal computer is simple, and one integrated circuit chip (cost about $8) adapts pulses to the wiring for the stepping motor.

One daunting feature of interfacing a computer to a bioreactor is obsolescence. A good workstation is expensive, but computers evolve so rapidly that a new model that can run the improved software is needed within two or three years. A computer purchased for several thousand dollars a few years ago is worth only a few hundred dollars today. However, we have found that old computers and their old data acquisition systems can be devoted to an individual bioreactor. Obsolete computers that were destined for the scrap heap serve in a master/slave arrangement. The master computer should be modern, but the slaves merely have to collect data and pass instructions from the master computer on to the control system of the bioreactor. When a slave computer needs repairing, we return it to the junk pile and replace it with another old computer. The analog-to-digital boards in old computers also wear out, but usually they fail one channel at a time. We move the wires from the sensors to a spare channel and continue on.

7
Conculsions and Predictions

Wieser and Satyro [59] have described current computing practices of chemical engineers and predicted the future. The bioprocess industries and their support network with universities and governmental organizations have benefited greatly from computers. You expect to see a computer at the desk of every scientist and engineer, and these computers are commonly networked. A period of each day is devoted to e-mail. The world wide web is used for retrieval of information. Our journals are moving to the Internet. Many bioprocesses are controlled by computers. When you have the proper permissions and are on the same network, you can look over the shoulder of the operators and follow a bioprocess in real time with your personal computer. It is easy to compare data from different runs, and there are software packages to aid your analysis.

I think that one feature that is lagging is "pages that do something." We are barely scratching the surface for better communication by adding interactive features to our documents. I am trying to do this with educational presentations, but the concept can be extended to all communications. Instead of long narrative documents, we can hyperlink from an index or abstract. The reader need not plow through the narration but can select the material of most interest. There can be links to supporting material, to spreadsheets, to archives of explanatory topics, to sketches, animations, video clips, and the like.

Programming skills of the typical student have deteriorated. Much of this can be blamed on commercial programs that render many tasks so easy. Our students can be expert with a word processing program, with spreadsheets, and with mathematical aids such as Maple or Matlab. They tend to master new programs quickly. However, they seem reluctant to write computer code. They would rather program their pocket calculator than write code. I have seen students write routines of great complexity using a spreadsheet program, never realizing that it would be much easier to write a program in BASIC or FORTRAN. I excuse this because modern spreadsheets have menu-driven schemes for developing spectacular graphs and charts. I have written code for so many years that I can get a useful graph almost as quickly in my own programs. The situation will probably get worse as the path of least resistance is to muddle through with the commercial programs that you know best.

Much of the exciting research of today will continue for several decades with incremental improvements and refinements. There are some areas, however, when current knowledge is meager and young investigators can make their marks. One is biological control mechanisms. There are some proposals for equations to handle mixed substrate systems, mixed culture systems, toxic nutrients or ingredients, preferential use of nutrients (diauxie), and the like. However, very little is known about how cells sense concentrations, how the information is processed, and how the controls function. A good example is the growth rate of cells. The Monod equation has been invaluable for describing what happens, but it says nothing about how the cellular controls develop the specific growth rate.

I predict that computerized presentations will more and more substitute for lectures. In fact, the lecture would be much more effective after preparing the students with a computer presentation. There is an analogy with a medical examination. The physician relies on tests before seeing the patient. A lecturer could perform better with students who have reviewed background material; the lecture could be modified if the teacher knew more about their strengths and weaknesses.

Fortunes have been made by locking users into an operating system. That may soon end as the Internet switches to its own operating system. At present, a file from the Internet must meet the specifications of an operating system. A much more simple operating system could handle display of files from the internet, and languages such as Java could execute faster by controlling the computer themselves instead of having another layer of executions demanded by an operating system. This could also expedite a revolution in the distribution of software. Instead of upgrading software yourself, you will pay for an Internet service that sends you the programs you need when you need them. These services would compete by having the newest and best software. I predict that commercial software will be layered so that you will not have to download all of a gigantic program. If you invoke a special feature, there might be a brief delay as this program element is downloaded and connected to its master program.

Biotechnologists are always among the first to embrace the advances in computers and software. As this continues, biotechnological education and research will make the bioprocess industries even more of a mainstay for improved living standards and for the global economy.

Appendix: Terminology for Process Dynamics and Control

- Block diagram
  The process is visualized as a series of interconnected functional blocks that correspond to process steps or to the controllers or the process sensors. A block acts on its input information and generates output. There may be several incoming signals and several output signals for a block. Typical signals in a bioprocess are concentrations, reaction rates, oxidation-reduction potential, pH, temperature, enzyme activities, and the like.

- Derivative or rate control
  Corrective action is based on the rate of change of the error. This is particularly suited to systems that can blow up. If the error is increasing rapidly, a derivative controller can institute a large amount of correction. It is uncommon to use solely derivative control. An industrial controller may combine proportional, integral, and derivative control, and the relative amounts of each are adjusted for a compromise between sluggish action and stability.
- Distance-velocity lag
  Delay of a signal for a period of time will give a graph of the inlet that is exactly the same as a graph of the outlet except for a displacement in time. For example, plug flow of a solution through a pipe may have negligible mixing or dispersion, thus an element of fluid may be assumed to traverse the pipe unchanged. Distance-velocity lag is a killer in terms of process control. Consider a sensor in a flow system where the sample takes time to travel from the bioprocess to the sensor. If the process variable is increasing and decreasing, the signal from the sensor can be out of phase so that the control action can be backwards from the actual process condition. The action that was supposed to make the process better can be in the wrong direction and make it worse. The way to reduce distance-velocity lag is to put sensors right in the vessel, to reduce distances between units, or to pump more rapidly through the connecting lines.
- Feedforward control
  The basic idea is to estimate the future error by calculating from the signals now. Feedforward control requires a model of the process. The model need not be exact because updated information from the process can correct for deficiencies. The better the model, the further into the future it can predict accurately.
  Practical controllers often mix feedforward control with feedback control. This can compensate for actual errors while allowing for estimated future errors. The models for feedforward control can be elegant systems of simultaneous differential equations or rather simple rules of thumb. Linguistic models supplemented by rules based on artificial intelligence can be quite powerful.
- Forcing functions
  The mathematics of process dynamics are derived from idealized signals. A *step* input changes to a new value suddenly. A step that returns to its starting value quickly is called a *pulse*. If the pulse has very short duration, it is an *impulse*. A *ramp* input changes linearly with time and is shown as a line sloping either upward or downward on a graph. A typical periodic input is a sine wave moving above and below its average value. Inputs to real processes may be approximated well by these idealized functions, or real inputs can be transformed by such methods as Fourier Series Analysis to a combination of sinusoidal inputs.
- Integral control
  The error (the difference between the actual and desired condition) is integrated and determines the amount of corrective action. Corrective action

thus accumulates to drive the offset to zero. Seldom is purely integral control used; a combination is common by which proportional control dominates and integral control minimizes the offset.

- Proportional control
  Corrective action is in direct proportion to the error. For example, if the bioprocess were far below the desired temperature, the heat would be turned to full on. As the system approached the set point, heat input would be decreased. Because there must be an error to get any corrective action, proportional control has a fundamental offset; there will always be a small error. Although the error can be decreased almost to zero by employing large gain (the multiplier in the transfer function), this may cause instability. Steady-state gain is the multiplier when the transients have died out.
- Set point
  The value selected as the target for control. The simplest controller is an on-off switch that supplies corrective action when a property of a system rises above or drops below its set point. The process response usually overshoots and oscillates above and below the set point. For sensitive processes, this variation may be unacceptable, and better control modes must be employed.
- Transfer function
  The response of a system to an input function depends upon the governing equations, and the expression that is used to derive the output from the input is called the *transfer function.*

References

1. Nagai S (1996) J Soc Ferm Bioeng 74:447
2. Bungay HR (1976) Biotechnol Bioeng 18:741
3. Bungay HR (1971) Process Biochem 6:38
4. Blackmore RS, Blome JS, Neway JO (1996) J Industr Microbiol 16:383
5. Sonnleitner B, Cheray A (1997) J Biotechnol 52:173
6. Franzen CJ, Albers E, Niklasson C (1996) Chem Eng Sci 51:3391
7. Schubert J, Simutis R, Dors M, Havlik I, Lübbert A (1994) J Biotechnol 35:51
8. Simutis R, Oliveira R, Manikowski M, Deazevedo SF, Lübbert A (1997) J Biotechnol 59:73
9. Ramakrishna R, Ramkrishna D, Konopka AE (1996) Biotechnol Bioeng 52:141
10. Narang A, Konopka A, Ramkrishna D (1997) Chem Eng Sci 52:2567
11. Shuler ML, Domach MM (1983) ACS Symp Series 207:93
12. Cui YQ, Okkerse WJ, Vanderlans RGJM, Luyben KCAM (1998) Biotechnol Bioeng 60:216
13. Sonnleitner B (1997) J Biotechnol 52:175
14. Galvanauskas V, Simutis R, Lubbert A (1997) Biotechnol Letters 19:1043
15. Galvanauskas V, Simutis R, Volk N, Lubbert A (1998) Bioprocess Eng 18:227
16. Omstead DR (1990) Computer control of fermentation processes. CRC Press, Boca Raton, FL
17. Rogers KR, Mulchandani AW, Zhou W (eds) (1995) Am Chem Soc Symposium Series 613
18. Rothen SA, Saner M, Meenakshisundaram S, Sonnleitner B, Fiechter A (1996) J Biotechnol 50:1
19. Agrawal P, Lim HC (1984) Adv Biochem Engr/Biotechnol 30:61
20. Edwards VH, Ko RC, Balogh SA (1972) Biotechnol Bioeng 14:939
21. Watson TG (1969) J Gen Microbiol 59:83
22. Fraleigh SP, Bungay HR (1986) J Gen Microbiol 132:2057

23. Levisauskas D, Simutis R, Borvitz D, Lubbert A (1996) Bioprocess Eng 15:145
24. Sakato K, Tanaka H (1990) Biotechnol Bioeng 40:904
25. Macias M, Caro I, Cantero D (1996) Chem Eng J/Biochem Eng J 62:183
26. Brass JM, Hoeks FWJMM, Rohner M (1997) J Biotechnol 59:63
27. Vanputten AB, Spitzenberger F, Kretzmer G, Hitzman B, Dors M, Simutis R, Schugerl K (1996) J Biotechnol 49:83
28. Chauvatcharin S, Siripatana C, Seki T, Takagi M, Yoshida T (1998) Biotechnol Bioeng 58:561
29. Preusting H, Noordover J, Simutis R, Lubbert A (1996) Chimia 50:416
30. Vanderwielen LAM, Vanbuel MJ, Straathof AJJ, Luyben KCAM (1997) Biocatal Biotrans 15:121
31. Gordillo MA, Sanz A, Sanchez A, Valero F, Montesinos JL, Lafuente J, Sola C (1998) Biotechnol Bioeng 60:156
32. Vanhamersveld EH, Vanderlans RGJM, Caulet PJC, Luyben KCAM (1998) Biotechnol Bioeng 57:330
33. Syu MJ, Chen BC (1998) Ind Eng Chem Res 37:3625
34. Simutis R, Oliveira R, Manikowski M, Deazevedo SF, Lubbert A (1997) J Biotechnol 59:73
35. Bulmer M, Clarkson AI, Titchenerhooker NJ, Dunnill P (1996) Bioprocess Eng 15:331
36. Zhou YH, Holwill ILJ, Titchenerhooker NJ (1997) Bioprocess Eng 16:367
37. Vanhalsema FED, Vanderwielen LAM, Luyben KCAM (1998) Ind Eng Chem Res 37:748
38. Stephanopoulos G, Han C (1996) Computers & Chem Eng 20:743
39. Diaz C, Dieu P, Feuillerat C, Lelong P, Salome M (1995) J Biotechnol 43:21
40. Diaz C, Dieu P, Feuillerat C, Lelong P, Salome M (1996) J Biotechnol 52:135
41. DeBernardez E, Dhurjati P, Lamb D (1986) Paper presented at Am Chem Soc National Meeting, Anaheim, CA
42. Sugeno M, Kang GT (1986) Fuzzy Sets Systems 18:329
43. Simutis R, Havlik I, Lubbert A (1992) J Biotechnol 24:211
44. Zhang X-C, Visala A, Halme A, Linko P (1994) J Biotechnol 37:1
45. Baugham DR, Liu YA (1995) Neural Networks in Bioprocessing and Chemical Engineering. Academic Press, New York
46. Vancan HJL, Hellinga C, Luyben KCAM, Heijnen JJ, Braake HABT (1996) AIChE J 42:3403
47. Vancan HJL, Tebraake HAB, Hellinga C, Luyben KCAM, Heijnen JJ (1997) Biotechnol Bioeng 54:549
48. Simutis R, Lubbert A (1997) Biotechnol Progr 13:479
49. Simutis R, Lubbert A (1997) J Biotechnol 52:245
50. Stoner DL, Miller KS, Fife DJ, Larsen ED, Tolle CR, Johnson JA (1998) Appl Envir Microbiol 64:4555
51. Shimizu H (1998) J Soc Ferm/Bioeng 76:338
52. Mateles RI, Ryu DY, Yasuda T (1965) Nature 208:263
53. Stephanopoulos G, Simpson TW (1997) Chem Eng Sci 52:2607
54. Varner J, Ramkrishna D (1998) Biotechnol Bioeng 58:282
55. Stephanopoulos G (1998) Biotechnol Bioeng 58:119
56. Simpson T, Shimizu H, Stephanopoulos G (1998) Biotechnol Bioeng 58:149
57. Boon M, Luyben KCAM, Heijnen JJ (1998) Hydrometallurgy 48:1
58. Stephanopoulos G, Locher G, Duff MJ, Kamimura R, Stephanopoulos G (1997) Biotechnol Bioeng 53:443
59. Weiser BS, Satyro MA (1998) Chem Eng Prog 98:51

Received June 1999

Automation of Industrial Bioprocesses

Walter Beyeler, Ettore DaPra, Kurt Schneider

PCS Process Control Systems AG, Werkstrasse 8, CH-8623 Wetzikon, Switzerland

E-mail: office@pas-ag.com

The dramatic development of new electronic devices within the last 25 years has had a substantial influence on the control and automation of industrial bioprocesses. Within this short period of time the method of controlling industrial bioprocesses has changed completely. In this paper, the authors will use a practical approach focusing on the industrial applications of automation systems. From the early attempts to use computers for the automation of biotechnological processes up to the modern process automation systems some milestones are highlighted. Special attention is given to the influence of Standards and Guidelines on the development of automation systems.

Keywords. Automation, Biotechnology, Process control, Computer control, Computer validation.

1 **Introduction** 140

1.1 Characteristics of Bioprocesses 140
1.2 Nature of Processes Automation 141
1.3 Structural Design 141
1.4 Problems of Hardware Architecture 143
1.5 Benefits of Bioprocesses Automation 144

2 **History of Automation of Industrial Bioprocesses** 144

2.1 General Considerations 144
2.2 Process Control Equipment for Sterile Conditions 144
2.3 Known and Predictable Process Behavior 145
2.4 Use of Automation Systems in Industrial Bioprocesses 146
2.5 Some Aspects on the Development of Programming Languages . . . 150

3 **Standards and Guidelines** 153

3.1 Why Standards? 153
3.2 Computer Directives History 153
3.3 Regulatory Directives 154
3.4 Non-regulatory Forums 156
3.5 Programming and Configuration Standards 157

4 **Outlook** 161

4.1 Integration of Enterprise Resource Planing (ERP) Systems into the Control Level 161

Advances in Biochemical Engineering/
Biotechnology, Vol. 70
Managing Editor: Th. Scheper

4.2 Ethernet-Based Device-Level Networks 162
4.3 A Trend in the Future? 162

References 163

1 Introduction

1.1 Characteristics of Bioprocesses

In the context of this study, the spectrum of bioprocesses is restricted to transformations of substances by microorganisms or cells in submersed cultures on an industrial scale to achieve one or more of the following goals: 1) degradation of complex substances into simple components, 2) synthesis of substances which may be accumulated in the microorganisms or excreted to the medium, 3) production of biomass from some nutrients. Usually the processes run in some kind of bioreactor to guarantee more or less homogeneous conditions and to perform the mass transfer of gaseous components creating the necessary turbulence.

Depending on the characteristics of the process and on the set goals, the process is carried out under sterile or non sterile conditions as a batch or continuous cultivation of one or more strains, the latter running in a stationary or, in some limits, in a non-stationary state. The different types of bioprocesses have different control requirements. A large-scale continuous culture for yeast production running in a stationary state is easily controlled by some parameters, such as temperature, pH, aeration and dilution rate; there is no demand for any logistics. In biological wastewater treatment running under continuous, non-stationary conditions, decisions are made depending on the highly variable input substances or on various maintenance requirements. Some type of logical reasoning has to be introduced into process control.

The two examples mentioned above would not justify a distinction between the control of a bioprocess and any other chemical process. However, regarding typical biological batch processes in the field of pharmaceutical production, running under rigorous sterile conditions, e.g., the production of insulin, interferon or vaccines, a high degree of complexity and special requirements justify a distinct discipline to describe the control of biotechnological processes. These processes consist normally of 5 main phases:

- equipment check,
- sterilization,
- cultivation,
- downstream,
- cleaning.

The plant itself is composed of dozens of piped vessels, hundreds of control loops, thousands of valves and many additional devices. The process is hardly in a stationary state and the setting of the entire periphery has to follow a recipe for the specific product. Exception states and alarm situations have to be mastered to protect human lives and costly equipment. Additionally, the validation of pharmaceutical processes and legal prescriptions requires sophisticated documentation describing all the details of each batch. In a modern biotechnological plant such processes run automatically.

1.2 Nature of Processes Automation

Today, automated devices are found in everyday life. From the standpoint of common sense, men are replaced by machines in automated systems. In earlier times automation was based on mechanics, as for example an electric piano, a machine fed by a program punched on paper rolls replaced the pianist. Rows of holes are scanned sequentially producing the time axis and depending on the row position an actuator was activated. Every activation turned the playing on or off. This simple example shows the basic characteristics of an automatic process: The system consists of an information processing machine communicating with a user (start/stop knobs), a process interface (driver of the piano hammers) and a program (punched paper rolls) evaluated stepwise. Depending on the automatic process the information processing is uni- or bidirectional. Therefore, the automated process can take into account process data inputs or establish a dialog with the user.

The basic element in automation is *the control loop* used to adjust an actual value to a given setpoint automatically. The feedback mechanism introduced by control loops is of primary importance for technical processes and, in industrial process control, many of these basic elements are needed. Like any automation, the control loop consists of a set of inputs, outputs and a program, however certainly on a level of lower complexity compared to the control of a whole process. The possibility to structure process control into different hierarchical levels is of primary importance, as will be explained in the next section.

1.3 Structural Design

Regarding the previous examples of pharmaceutical production, processes are composed of thousand of devices, the implementation of automation is only possible on the basis of a well-established structure using a system of one or more computers including process and user interfaces, program libraries and data bases. The list of process interfaces defines all of the I/O-channels available between the process control system and process field. The operations that can be accomplished have to be functionally described at least in three hierarchical levels:

Control Level 1. At the lowest level, I/O-Channels are grouped together to control loops, as for example the control of temperature, pressure, gas and fluid flow or pH-value. Control is realized by looping step-function sequences representing in most cases a PID-algorithm with a constant scanning interval. Therefore the computer has to guarantee a deterministic time behavior otherwise the PID-algorithms will not work correctly. The control loops can be switched on and off by the operator using the user interface. If the user interface allows the setting of all the outputs and displays all the inputs, the process could be run "by hand" switching on and off the respective control loops or output devices and adjusting the setpoints to the required values. This level is called DDC-level (Direct Digital Control).

Control Level 2. The next higher level defines logistic autonomous process units with the capability of performing a set of operations. Process units may be represented in a hardware structure, like bioreactors, transfer pipes, medium tanks, downstream equipment etc., or they can represent an abstract logical unit, like a scheduler servicing CIP-requirements of other process units. The same operation can run with different sets of parameters. By selecting the right parameter-set, the operation can be adapted to a specific product. An operation may start only under some defined conditions. Locking mechanisms are based on the state of the process unit, and exception routines are required to avoid disastrous situations. According to the functional description the operation has to document its own execution. At level 2, the requirement concerning the time behavior is much less critical compared to the DDC-level, access to mass storage devices and network communications should, however, not disturb the operation's execution. With a combination of level 1 and 2, processes of medium complexity can be automated. The user starts the needed operations, in the right sequence following a written recipe or Standard Operation Procedure (SOP). For a continuous process typically only a few operations have to be started. However in a batch process, the user has to control a large number of sequential operations, starting each operation and waiting for its completion. As this is very time-consuming, manual procedures are not satisfactory for complex production processes and a third level of control and automation is needed.

Control Level 3. The third level represents a scheduler starting operations according to a recipe. Formally, the recipe may be represented by a graph, where each node corresponds to an operation waiting for its start after the termination or completion of one or more running operations. If an operation fails, the corresponding process unit will not reach the end state that the scheduler is waiting for. A simple strategy to handle such exception states is based on a manual restart of the failed operation. As soon the corresponding process unit reaches the correct end state the scheduler will continue and start the next operation.

This short description of the three levels illustrates that by proper structuring the automation of complex bioprocess can be managed. There are still, however, many more important problems left to consider. Some of these are universal in computer applications, such as the problem of creating an optimal "Human-Machine-Interface".

It is obvious that a successful implementation of a control system depends on an adequate documentation comprised of at least the user requirements, functional descriptions, program code, and test procedures. This implies that a rigorous quality management system should accompany the implementation as it is now defined for the pharmaceutical industry in the Suppliers-Guide from the GAMP-Forum [1]. Following these guidelines a firm base for validation is achieved.

1.4 Problems of Hardware Architecture

The structural principles outlined above can be realized with many different hardware architectures, computers, and process periphery. The architectures differ mainly in the degree of centralization.

The process interfaces can be connected directly to the computer and the wiring to the transmitters may be realized in the form of a star topology. On the other hand, a decentralized solution could consist of a fieldbus system with interfaces distributed throughout the whole process area and connected serially to the fieldbus. In a typical fieldbus controller; the bus topology is mirrored in the controller's memory and contains the actual value of each device updated in time intervals of some milliseconds. The computer no longer has to directly access the interfaces. Today, fieldbus systems are widely accepted: the cabling is transparent and allows for easy maintenance, the interfaces are close to the equipment and decoupled from the computer, the computer itself has only to read from and write into memory cells inside the fieldbus controller to exchange data with the process periphery. Such systems are extremely highly reliable and have sophisticated error-detecting facilities built in.

The question concerning the "right" computer system architecture is discussed among suppliers and users in a controversial manner. In the past, process computers depended on low performance processors with low storage capacities (memory, disk devices). Consequently the development of process control systems led to distributed systems with a lot of small computers controlled by one or more master computers. However the performance and capacity of today's computer systems and the availability of powerful real-time operating systems allow the realization of process control systems of any degree of complexity in one single computer. For today's pharmaceutical applications, reliability is probably the most important requirement. But it is very difficult to compare the reliability of a multi-component distributed system requiring a considerable amount of network traffic to a single component system without a network to other process computers. Maybe nature itself has given the answer to this problem. During their evolution biological systems ended with a centralized computer system, and a decentralized peripheral system. Perhaps a process control system should evolve in the same manner. Moreover a centralized computer system consisting of two redundant computers and a redundant fieldbus system seems to be an optimum solution. But as the main requirement for industrial applications, the systems should be as simple and robust as possible. Furthermore many interesting developments have been developed in the academic world (sophisticated controllers, process optimization, application of Artificial Intelligence (AI) etc.) are still too sensitive to survive in the daily life of an industrial production process.

1.5 Benefits of Bioprocesses Automation

- Automation makes it possible to run bioprocesses of any degree of complexity.
- The recipe-controlled batch guarantees a product of constant quality accompanied by the necessary documentation of the production process. At any time, the batch corresponding to an individual product can be traced back to its origin. Accurate documentation is needed to fulfill legal or validation requirements.
- Automation increases the reliability because the operator is supported by the automation system (check lists, alarm messages, help libraries).
- The safety for humans and materials during the production process is substantially increased because an automation system checks the critical parameters continuously. In addition the system is capable of handling failures according to defined exception routines.
- The economics are improved as time and personnel are saved.

2 History of Automation of Industrial Bioprocesses

2.1 General Considerations

The term automation may be ambiguous. The meaning may be automatic control (e.g. control of temperature), but automation, as it is understood by the authors is control on a higher level and involves sequences of tasks based on a schedule or a series of events as defined by Singh et al. 1990 [2]. In this historic retrospect, we will mainly focus on automatic operation.

The history of the implementation of automatic operations into biotechnological processes may be as old as the history of biotechnology. It was probably always man's intention to replace repetitive operations by machines. However modern automation developed parallel to the development of the technologies and equipment needed for process automation. It may be generally observed that in industrial biotechnological processes, technologies and equipment available on the market have been adapted to the special requirements and implemented. There are only a few exceptions where control and automation equipment were specially designed and developed for biotechnological processes. It is only the requirement for sterile operation of biotechnological processes that asked for special developments. In Biotechnology, sensors and actuators have to be able to withstand sterilization conditions and as sensors and actuators are an essential part of any automation system a few words on the history of these components will be given in the first part of this historical consideration.

In theory, process automation is only possible if the process itself is known and the process behavior can be predicted at any time. Therefore, besides the availability of control and automation equipment, it is an important necessity that these two requirements be thoroughly understood and respected. For biotechnological processes these may still be the weakest points in the attempt to automate bioprocesses. A few considerations to these problems will be mentioned in part two of this retrospective.

Automation needs equipment capable of acting according to preset patterns or algorithms but there are also logical components needed to automate a process. The development of logical devices from simple relays to modern process computers had a continuously strong influence on the development of automation systems. In Sect. 3, this influence of new electronic control equipment on process automation in biotechnology will be reviewed.

To summarize, successful automation of industrial bioprocesses is only possible if 1. the equipment (sensors and actuators) for sterile processes is available, 2. if the process in its basic behavior is known and predictable and 3. if the controllers with the needed algorithms are available. An attempt to illustrate the history of bioprocess automation has to take these points into consideration. The following are prerequisites for bioprocess automation:

- Availability of Field equipment suitable for sterile operation (sensors and actuators);
- Known and predictable Process Behavior;
- Availability of reliable Automation Systems (Computer Systems and Software).

2.2 Process Control Equipment for Sterile Conditions

Even in the first industrial bioprocesses for the production of antibiotics in the late 1940s, automatic features such as control loops for temperature and impeller speed had been implemented. As at this early stage of industrial biotechnology, sterility was not absolutely necessary, techniques developed for other applications could easily be adapted to bioprocesses. However, with the requirement for sterile conditions the need for special sensors and actuators arose. As an example, the measurement and control of the pH-value should be mentioned. In this context, the development of the first sterilizable pH-electrode by Fiechter et al. 1964 [3] has to be considered as an important milestone in the history of bioprocess automation. As a result of this pioneering work, one of the most important parameters for biological reactions could now be measured and controlled under sterile conditions. Of comparable importance on the way to an automated bioprocess was the development of membrane valves capable of operating under sterile conditions as introduced by several equipment manufacturers in the 1970s. In the scientific literature, these valves have not been considered worth mentioning. However, only these valves allow for an interaction with the process while maintaining sterility and it is hard to imagine any modern biotechnological processes without them.

These two examples may be representative of the importance of suitable field equipment for the automation of bioprocesses. They are mentioned to illustrate that process automation comprises not only electronics and computers. Without sensors and actuators, even the most sophisticated computer system would be useless.

2.3
Known and Predictable Process Behavior

In theory, automation is only possible if the process behavior is known and predictable at any time. Although knowledge about biological reactions has increased immensely during the last decades, it never would allow us to interpret the extremely complex behavior of biological systems. With its huge variability, a biological process is not predictable. The on-line measurements do not contribute much to overcome this lack of knowledge as there are still only a few exceptions known where biological quantities such as biomass, products, intermediates or substrate can be measured on-line in an industrial environment. Consequently far more than 90% of the scientific publications about "Bioprocess Control" focus on these problems. New analytical procedures, new sensors, process and control models, optimization and its implementation into control strategies, dominate the scientific literature. As soon as the first mini-computers appeared on the market, biotechnologists all over the world used computers to calculate process parameters based on various process models. With the appearance of personal computers this tendency even increased. The recently published Proceedings of the 7th IFAC International Conference held in Osaka from 31st May to 4th June, 1998 [4] gives an excellent overview on the present status of research and development activities in this field. It is not the authors' intention to review these scientific research and development activities. These are certainly of great importance for the understanding of biological reaction systems and they may be used once in future automation systems. Up to now, none of these models have been implemented into real industrial production processes. Industrial biotechnology seems to operate pragmatically and does not care about the lack of basic knowledge about biological reaction systems. By dividing the whole process into smaller process units with known behavior, by taking off-line data and experiences into the control concept and with a combination of automation with manual interactions, a high degree of automation can be achieved despite the fact that the detailed behavior of the process itself is not known. This proves that the limiting factors for successful automation of biotechnological processes are more technical rather than biological.

2.4
Use of Automation Systems in Industrial Bioprocesses

In addition to the previously mentioned prerequisite of the availability of field instrumentation and process knowledge, successful process automation requires a logical device to control the operation according to a preset. Even in the very first industrial bioprocesses for antibiotic production, the technical equipment to control and to automate these processes was available. Analog controllers mostly configured for PID-Operation (Proportional-Integral-Derivative) had been used to solve all of the control tasks. Logical elements such as timers and relays had been allowed to fulfill the automation requirements. As these hard-wired techniques were very difficult to realize and to maintain, automatic operation was normally limited to defined small standard operations such as sterilization, product harvest or media transfer. Automation of complete production plants or even recipe handling was not feasible with this hard-wired techniques.

This changed completely when, in late 1960, a newly designed solid-state controller was introduced into the process control market. This new device, called a programmable logic controller (PLC), not only replaced the relay logic controllers, but more importantly offered new functionality not yet realized with conventional analog controllers. These PLCs were quickly implemented into biotechnological plants; at the beginning just replacing the conventional relay logic. All leading plant manufacturers at that time realized standard operations with PLCs.

The PLC is functionally divided into four parts: the input, the output, the logic unit and the memory unit. This basic principle has remained valid until now although the PLC has become much more powerful (more memory, speed) and flexible (more functionality) in the last decades. Still, PLCs are widespread in biotechnological plants and are used to do much more than simple control sequential actions. Whereas single stand-alone equipment (such as a centrifuge or a filtration unit) is relatively simple to automate with a PLC, the automation of complete plants comprising several bioreactors, tanks, up- and downstream equipment is not within the PLCs reach. The immense effort to coordinate the actions of single PLCs to handle a recipe that requires multiple devices and various equipment may end in an immense traffic jam in communication. In the extreme, each PLC has to mirror the status of all the other PLCs in the same production unit. Consequently, today process computers are replacing PLCs more and more.

In the 1960s, the general-purpose digital computer was brought to the market and soon after also applied to biotechnological applications. In contrast to PLCs, these general-purpose computers offered a complete versatility irrespective of the application. The functionality was defined by software alone. Additional features such as mass storage, communication networks, visualization devices as well as an operating system controlling the interactions of the different modules, were now available. The initial differences, mainly based on performance and price, between micro-, mini- and mainframe computers has decreased more and more over the last two decades. Biotechnological companies and research institutions recognized very quickly the great potential of these universal computers and used them to acquire and to store data, to control process parameters, and to automate operation sequences. Furthermore due to the high calculating power of these machines, on-line process modeling became possible. A favorite among the computers used in those early days was the computer series PDP 8 to 11 (Programmed Data Processor from Digital Equipment Corporation).

No publications have been found showing the industrial use of these minicomputers in the early days. Therefore the authors contacted all leading biotechnological companies as well as manufacturers of biotechnological equipment to get information on the early use of computers for process automation. Unfortunately only three answers have been received from more than

50 requests sent out. On a follow-up by phone, the authors received mainly the same answer: all relevant data had disappeared. It seems that during these dynamic developments of the last three decades, nobody considered the historical value of data and equipment. Therefore, this review is mainly based on the personal experiences of the authors and may not correctly represent the whole situation.

New Brunswick Scientific Co. Inc., probably the first plant manufacturer offering bioprocess systems controlled by minicomputers, provided us with some pictures of the early days of computer applications in biotechnology and we feel it worthy enough to publish these pictures as historical documents (Fig. 1). Looking at these pictures it is hard to believe that there are only 25 years of development between the system shown and a modern process automation system used today. The computer system to control a relatively simple pilot plant needed a complete room, which also had to be air-conditioned. Memory at that time was limited and the programmers were forced to optimize their code in order to save space. A very efficient real-time operating system organized the available memory of 128 kbytes in a way that much larger applications could be executed successfully. The "Human-Machine-Interface" was at that time a video terminal with a keyboard and the information displayed was completely text based. No on-line graphics display was available at that time.

This period characterized by the use of the PDP11, may be considered as a real milestone not only for the development of automation systems for biotechnological applications, but also for the general understanding and further development of the whole biotechnological industry. It initiated a remarkable change in the mostly biology oriented biotechnology of that time. From that time on, natural science was definitively influenced more and more by engineering sciences and biologists had to learn to communicate with engineers. Biologists had been forced to describe the "Art of Fermentation" and to convert their experiences into Bits and Bytes. The way of looking at a bioprocess had completely changed.

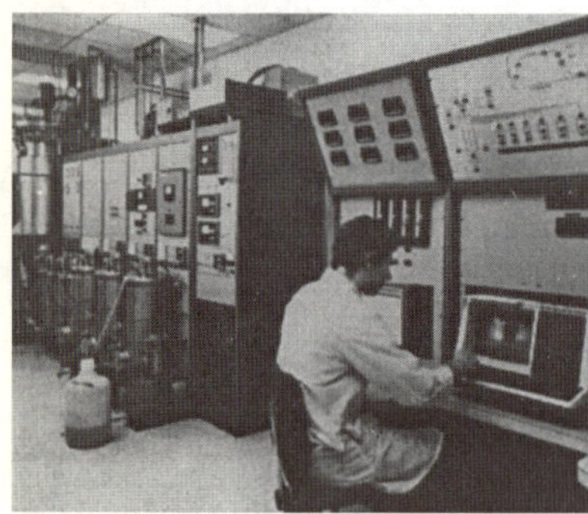

Fig. 1. Control of a biotechnological pilot plant in 1978 by a PDP11 computer system. With courtesy of New Brunswick Scientific Co. Inc.

The period of the PDP11 was only of short duration. New computer generations arose with much more powerful components (e.g. HP-1000 series from Hewlett Packard, Honeywell 4500 or DEC Vax Series) were soon brought onto market. Fortunately, this development of new computers went in parallel with a generally prosperous growth of the pharmaceutical companies. Some of these companies started strong investments into biotechnology and during the late 1970s and early 1980s many new pilot plants and production facilities were built. All of these plants had been equipped with the latest computer products available at that time. Reports have been published on installations at Elli Lilly and Company [5], Genetic-Institute [6], Merck Sharp and Dohme [7, 8], Schering-Plough [2] and Smith Kline [9]. All of these biotech companies elaborated on the use of computers to control and to automate their bioprocesses. They demonstrated not only in different ways the principle functioning of computerized biotechnological plants, but added substantial new features to the systems. The focus of that time was no longer on just the control and automation but also on the evaluation and management of process data. Process computers were networked, supervisory systems implemented and communication with company information systems was realized. The networked bioprocess was born. Simple graphic displays started to replace the text based "Human-Machine-Interface" and made the system accessible to people who were not computer engineers.

Common to all these installations was a very high investment in equipment and manpower needed to keep the system running and it is obvious that this restricted the numbers of possible users. With the introduction of the Personal Computer (PC), in 1977 by Apple, computers were made more affordable and available to a broader range of companies and institutions. The age of PC-controlled processes was born. In research institutes all over the world, activities went on to couple bioreactors with PCs. Pioneering works on the use of PCs for on-line calculations of biological process parameters were published by Röhr et al. 1978 [10]. One of the first PC-based bioprocess systems commercially available on the market was developed by B. Braun Biotech International GmbH, Melsungen [11] 1982. Based on an HP-85 desk-top PC, set-point-control (SPC), data acquisition and visualization, as well as automatic sterilization and even some simple on-line calculations of biological properties such as RQ-value had been realized. It is not known to the authors how many of these systems were sold. It may have been only a few units because the technical improvement of PCs was so fast that shortly after the appearance of a new system on the market, the products themselves were already obsolete. However, historically this example may be of considerable importance as it stood at the beginning of a new trend for equipment manufacturers. Since that time, all of them have had to develop and maintain PC-based control systems. Still today in smaller installations PCs are used to control and automate bioprocesses. Together with the very powerful control and automation software available now, the PC has developed into a wide spread automation device and has replaced more and more the traditionally applied PLCs.

Attempts to automate complete industrial biotechnological production plants have only been undertaken since the late 1970s and were initiated by the

Table 1. Functional levels of process automation systems

Functional level	Main functions
Data Communication Highway (ETHERNET)	Communication with further information systems
Supervisory computers	Plant automation, recipes handling, data bases
Operator Interfaces	Operator interactions with the process
Front-end Process Computer	Real-time control and automation of single process units
Sensor/Actuator Fieldbus	Field data handling communication with process computer
Process	Process 1 to n with Sensors and Actuators

powerful process computers which appeared on the market. In 1975, Honeywell introduced the first process computer system (TDC2000) and shortly after this system was used in a biotechnology plant. With this system a standard was set. All leading automation companies such as Siemens, ABB, Fisher-Rosemount, Foxboro, just to mention a few, soon brought comparable systems onto the market.

In parallel with the development of new process computers, different companies elaborated new concepts for field instrumentation. Instead of the traditional centralized communication of sensors and actuators with the computer system, fieldbus systems for decentralized communication with sensors and actuators were developed. The biotechnological industry soon realized the great advantages of this fieldbus concept and implemented it in plant automation systems. Whereas in research and laboratory environments, the CAN-Bus is often used [12], in industrial biotechnological plants, mainly Profibus and Interbus-S can be found.

The general concept of a modern up to date automation system for biotechnological processes comprises many of the different functional levels shown in Table 1.

Depending on the computer hardware applied, different functional levels may be implemented into one computer system.

2.5 Some Aspects on the Development of Programming Languages

For the very first computer based systems, the programs had to be written by the users themselves. The programming language at that time was mostly FORTRAN. Additionally, some time-critical tasks had to be written in Assembler (a CPU-architecture-dependent low-level language). Although there are still some FORTRAN programs around, FORTRAN was replaced more and more by languages like Pascal, C and later C++. Together with the PCs, the BASIC language was developed and most PC-based applications had been written in this language. It soon turned out that the "Spaghetti-Code" of BASIC programs was not appropriate for complex applications. BASIC disappeared almost completely and is replaced by either Visual Basic or commercially available control and automation packages today.

Writing the application software was the crucial point in all automation projects and there were practically no projects where the costs and time for software development had not been underestimated. At the beginning of the computerized age, the industrial users set up their own automation groups and developed the software in-house with the important disadvantage that all the efforts made could normally only be applied once. This may be the reason that more and more specialized automation engineering companies emerged. These companies first started with their own software developments but more and more went over to base their application on generally available control and automation software packages.

Furthermore, equipment and plant manufacturers started to provide their equipment with their own automation software packages such as B. Braun Biotech International GmbH, Melsungen does with the MFCS for bioreactor automation [13] or New Brunswick Scientific Co with the AFS-BioCommand. These Programs may serve quite well to automate the equipment of these companies but can hardly be applied to other manufacturers equipment or to automate complete plants.

The appearance of universal configurable software packages for process control such as Genesis, Lab-View and others may help to minimize investment in time and money. Examples of the successful use of such commercially available automation packages have been published by different authors [14, 15]. All of these packages offer graphically oriented configuration tools and are relatively easy to handle, also by non-specialists. In addition, such software packages offer the advantage that the user normally does not have to care about future upgrades and compatibility with new computer or operating systems. It may be assumed that the software company takes care of this. A disadvantage may be the dependency on the software supplier. If a special need of the user is not part of the package it is nearly impossible to fulfill this need.

Large automation companies offer software solutions together with their computer hardware products. All of these general automation software packages can be used for biotechnological applications. However, the configuration of the software needs automation specialists to convert the functional process description into program code. Very intense communication between the biotechnologically oriented user and the general automation engineer is essential for a successful realization. There are projects known to the authors where famous automation companies failed to automate biotechnological processes, not because of a general hardware deficiency but as a consequence of a communication problem between the biologically oriented users and the technologically oriented automation engineers.

Although the history of automation of industrial biotechnology is only about 25 years old, it is extremely difficult to structure and to weight the data. The history does not comprise single events that can be chronologically ordered but consists of complex interactions of various parameters (Table 2). The development of new computer products was certainly one of the driving forces, but it was not the only one. Of similar importance may be the biotechnological in-

Table 2. The evolution of computers and process automation systems for industrial biotechnological application

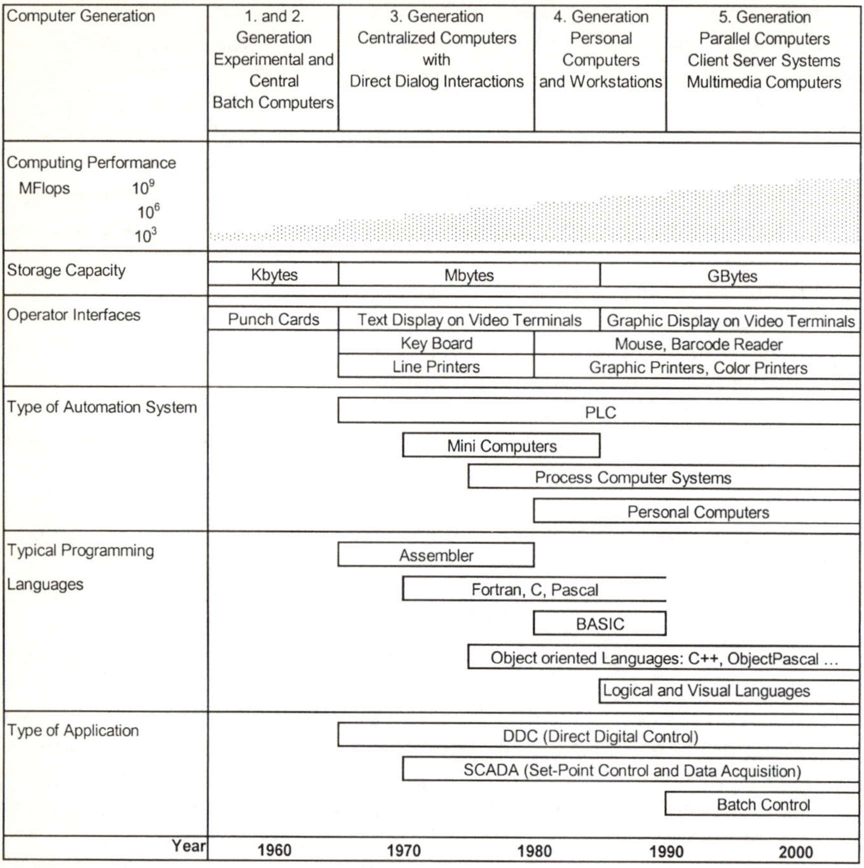

Computer Generation	1. and 2. Generation Experimental and Central Batch Computers	3. Generation Centralized Computers with Direct Dialog Interactions	4. Generation Personal Computers and Workstations	5. Generation Parallel Computers Client Server Systems Multimedia Computers
Computing Performance MFlops 10^9 10^6 10^3				
Storage Capacity	Kbytes	Mbytes		GBytes
Operator Interfaces	Punch Cards	Text Display on Video Terminals		Graphic Display on Video Terminals
		Key Board	Mouse, Barcode Reader	
		Line Printers	Graphic Printers, Color Printers	
Type of Automation System		PLC		
		Mini Computers		
		Process Computer Systems		
			Personal Computers	
Typical Programming Languages		Assembler		
		Fortran, C, Pascal		
			BASIC	
		Object oriented Languages: C++, ObjectPascal ...		
			Logical and Visual Languages	
Type of Application		DDC (Direct Digital Control)		
		SCADA (Set-Point Control and Data Acquisition)		
				Batch Control
Year	1960	1970	1980	1990 2000

dustry itself with its special requirements for process automation providing pressure for new system developments. The extremely dynamic computer market, the rapid appearance of new software developments and last but not least the dynamics of the biotechnological industry itself, makes it difficult to draw conclusions for future tendencies. The trend for future computer developments will still go on: *smaller, faster and more powerful!* On the other hand there is the industrial need for *reliability and continuity*. This will slow down the integration of new hardware into process automation. In addition the increasingly stricter regulations and *standards* for automation systems for the pharmaceutical industry will greatly influence future developments. It may therefore be concluded that the generation time for industrial automation system will still be much larger than the generation time for computer hardware.

3
Standards and Guidelines

3.1
Why Standards?

Before proceeding into details of the different standards involved in process control automation, there is a brief discussion about standards themselves. Why are standards needed? Who determines that they are standards? Who enforces compliance with standards?

Standards are necessary in order to engage in commerce of any type. Component parts for any kinds of "final product" are manufactured at many different locations by different vendors. Each individual contributor involved must design, manufacture and test his components according to the specifications he received. In the "final product", assemblies of different parts must fit and work together as a whole as specified by the customer, even though the individual contributors may never have even met. Without common agreement, common reference and common understanding, such a procedure would not be possible. Therefore, national standards have been developed in all major industrial nations. The units of the standards might differ from nation to nation; however, conversions between them are possible as long as everyone concerned understands the standards themselves and as long as those standards selected remain constant. The need for international standards has been recognized by different national standard organizations and has given birth to international non-governmental bodies such as the International Electrotechnical Commission (IEC) founded in 1906 and the International Organization for Standardization (ISO) established as an organization in 1947. National or international vendor or user-groups set up other standards. To go into detail concerning the specifics of the different organizations and standards is beyond the scope of this paper. However, some standards related to the process control systems will be detailed further in the next sections.

3.2
Computer Directives History

Regulatory authorities around the world have come to recognize the importance of electronic data systems used in the research, development and manufacture of products and services coming under their jurisdiction. For more than a decade, authorities have been issuing guidance documents for the industries and to assist their inspectors in how to review systems for Good Practice compliance. The percentage of computer content in regulated activities has been growing rapidly since 1980. With this growth in the use of computers has come a parallel growth in regulatory guidance for computer validation to assure the quality of electronic data handling and electronic process controls.

With the growth in regulatory directives, forums were founded to promote the understanding of the regulation and use of computer and control systems

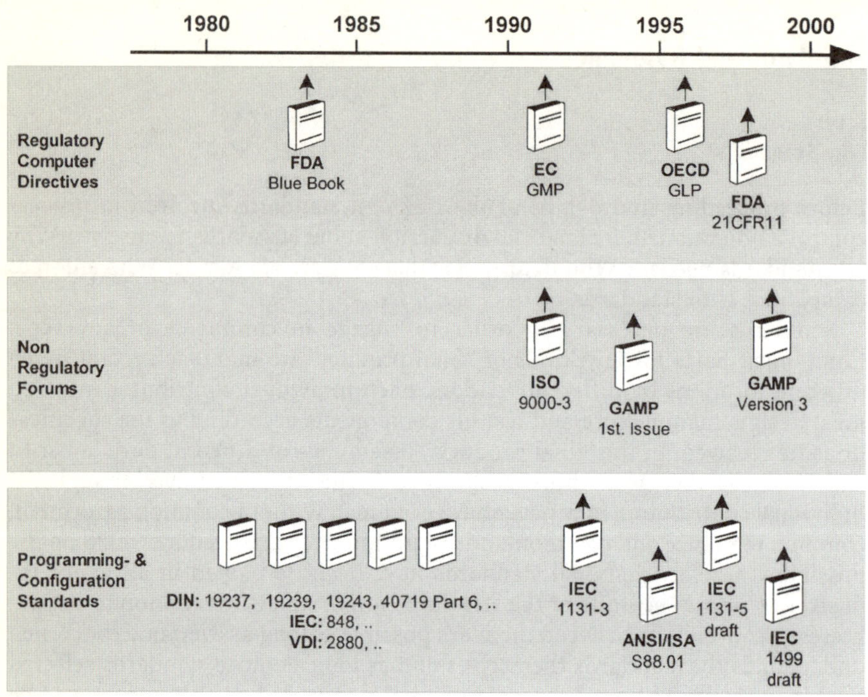

Fig. 2. History of computer directives and standards

within the pharmaceutical manufacturing industry. Different "non-regulatory" standards were also established by such forums.

Independent of regulatory issues, the industries and the national- and international committees agreed to standards defining programming and configuration issues in computer control. Figure 2 illustrates the history of process control computer directives and standards. These "documentary milestones" detailed further in the following chapters represent the personal point of view of the authors, as the result of more than twenty years of involvement in the field of automation of biotechnological applications.

3.3 Regulatory Directives

1983: The Blue Book. The first document published in 1983 became known by the color of its cover – the Blue Book [16]. Its focus is on computerized systems used in drug manufacturing and instructs inspectors of the US Food and Drug Administration in computer characteristics and how to inspect systems under the regulations for Good Manufacturing Practices (GMP). The purpose of the guide is to provide the field investigator with a framework upon which to build an inspection of drug establishments which utilize computer systems. The document was not intended to spell out how to conduct a CGMP drug inspection or set forth reporting requirements, but rather what aspects of computerized systems to address during such inspections and suggestions on how to address the systems. The guide discusses some potential problem areas in application of computer systems, provides inspectional guidance, and includes a glossary of terms the investigator should be aware of prior to performing the inspection.

The blue book triggered mystification among the manufacturers and the suppliers of automation systems. FDA validation advanced to a key issue in each automation project, even in Europe. The fact was that most people had problems in transforming this guidance into their own concise quality and validation plan. The subsequent issue of supplemental compliance policy guidelines [17–23] and the Pharmaceutical Manufacturers Association (PMA) Computer System Validation Committee (CSVC) concept paper [24] are the basis for today's computer system validation practice.

1991: GMP – Europe. The European Community adopted directives laying down principles and guidelines of good manufacturing practice (GMP) for medicinal products. The Guide to Good Manufacturing Practice [1] is used to assess applications for manufacturing authorizations and is a basis for inspection of manufacturers of medicinal products. The first edition of the guide was published in 1989, including an annex on the manufacture of sterile medicinal products. The second edition was published in January 1992; including the Commission Directives 91/356 of 13th June 1991 [25] and 91/412 of 23rd July 1991 [26] laying down the principles and guidelines on good manufacturing practice for medicinal products for human use as well as for veterinary medicinal products. The second edition also included 12 additional annexes. The basic requirements in the main guide have not been modified, but 14 annexes on the manufacture of medicinal products were included in the 1998 edition.

Annex 11 [27] to this guide gives a concise view of good computer practice in nineteen paragraphs of directives covering the training and experience of personnel, the extent of validation throughout the whole life cycle of a system, and operational guidelines for system development, deployment, risk management and supplier contracts. Annex 11 reflects four major themes:

1. *Management Control:* Does the company apply System Life Cycle activities?
2. *Data Integrity:* Is electronic data handled properly?
3. *Reliability:* Is the systems continuous performance reliable?
4. *Auditability:* Is there documented evidence for all above topics?

1995: OECD GLP Consensus – Europe, Japan & US. The Organization for Economic Development and Cooperation (OECD), which includes Europe, Japan and US published a consensus document for applying GLP (Good Laboratory Practice) principles to computerized systems [28]. The OECD document contains more as Annex 11, and also adds other items of concern such as the definition of raw data and retrospective evaluation. A more detailed comparison of validation guidance has been issued in a white paper by the Digital Equipment Corporation (DEC) [29].

1997: 21 CFR 11. In 1997 the FDA issued regulations that provide criteria for acceptance by the FDA, under certain circumstances, of electronic records, electronic signatures, and hand-written signatures executed to electronic records as equivalent to paper records and hand-written signatures executed on paper [30]. These regulations, which apply to all FDA program areas, are intended to permit the widest possible use of electronic technology, compatible with FDAs responsibility to promote and protect public health.

3.4 Non-regulatory Forums

Several groups around the world, including the Instrument Society of America (ISA), the International Electrotechnical Commission (IEC), the International Organization for Standardization (ISO), the GAMP group, Germany's Normenarbeitsgemeinschaft Mess- und Regelungstechnik der Chemischen Industrie (NAMUR) and others are working on improving and standardizing different fields of process automation.

1991: ISO 9000-3. ISO prepared the 9000-3 guideline [31] in order to help organizations to apply the ISO 9001 standard to computer software development, supply, installation and maintenance. The ISO 9000-3:1997 edition is actually an expanded version of the old ISO 9001:1994 standard. ISO has simply copied the old text from ISO 9001 [32] and pasted it into the new version of ISO 9000-3, and then added some new text that refers only to software.

The spirit of ISO 9000 is that customer and suppliers should work together to guarantee consistent quality standards and to have procedures in place so that when the quality standards move away from those the customer requires, corrections are made. However, ISO 9000 allows the supplier to have procedures in place, without really taking into account the customers needs. ISO 9000 is an accreditation system, which lacks technical depth.

1994: The GAMP guide. With the increased penetration into, and complexity of automated systems in pharmaceutical manufacturing, the focus on such systems has increased. With it comes the need to improve the understanding of the regulations and their interpretation. Better communication was required, not only within the pharmaceutical industry, but also with its suppliers. Thus an informal group, the UK Pharmaceutical Industry Computer Systems Validation Forum (now known as the GAMP Forum) was set up to promote that understanding. Draft guidelines entitled "Good Automated Manufacturing Practice (GAMP)" were created by the forum led by this forum. Its remit was:

1. To devise a draft set of guidelines for suppliers of automated systems to the pharmaceutical manufacturing industry.
2. To take account of the requirements of both the European and North American regulatory bodies.
3. To make use of existing internationally recognized standards where appropriate.
4. To consult with the Medicines Control Agency in the UK.

The first draft was approved by the Forum and made available for comment from the suppliers of automated systems to the industry and other interested parties in 1994. The current version, GAMP 3 [1], enables compliance with the requirements of the European Community as explained in the European Guide to GMP including Annex 11 Computerized Systems", and in particular item 2 (validation and the Life-cycle) and item 5 (software quality assurance) for system development and validation. Systems designed and implemented in accordance with this guide should also meet all the other requirements of the EC guide. This approach should also be acceptable to the FDA.

The Guide is built around a formal management system for the development, supply and maintenance of automated systems by a supplier. Adherence to this management system by the supplier will provide both the system and sufficient documentary records to enable the complete system to be accepted and validated by the user. The formal acceptance of an automated system and its documentation by the user is an integral part of the validation of that system. This formal acceptance by the user is a key issue. Due to the fact, that this documentary evidence is required by the pharmaceutical companies as an integral part of validation, it is defined as: "Establishing documented evidence which provides a high degree of assurance that a specific process will consistently produce a product meeting its pre-determined specifications and quality attributes".

The actual GAMP 3 guide is like a fill-in-the-blanks manual for the supplier and the client and is designed for prospective validation of new computer systems. It helps focus on all the details needed for specification and subsequent verification of a computerized system's functionality. It also helps to allocate validation support responsibilities between supplier and customer. The guide consists of a Supplier Guide section, a User Guide section and a Best Practice Guide section.

The Supplier Guide section defines a management system model with life cycle and supporting activities for the supplier.

The User Guide section outlines the different user responsibilities and illustrates a possible validation strategy for different types of software products. The Best Practice Guide section details the validation procedures for Information and Control Systems and Systems.

Conclusions. The GAMP guide represents an excellent basis for a modern automation project management. But one question remains: Who will bear the costs? The investments in the development (specification) process are considerable by means of time and money. Tight time schedules and project budgets increase the problem. The consciousness, however, must be present, that the invested money will be paid back over the lifetime of the product.

3.5 Programming & Configuration Standards

1993: IEC 1131–3. Automation system designers are often required to use programmable controllers from various manufacturers in different systems, or even in the same system. However, the hardware of programmable controllers from different manufacturers may have very little in common. Historically, this has resulted in significant differences in the elements and methods of programming the software as well, to the existing standards and regulations (DIN 19239, DIN 19237, DIN 40719 Part 6 and VDI 2880). This has led to the development of manufacturer-specific programming and debugging tools, which often carry very specialized software for programming, testing and maintaining particular controller "families".

An international committee was set up to tackle this problem and produced the world-wide standard IEC 1131 [33], the third part of which deals with the PLC programming languages. A major goal of IEC 1131-3 is to remove barriers to the understanding and application of programmable controllers. Thus, IEC 1131-3 introduces numerous facilities to support the advantages of PLCs, even if controllers of different vendors are concerned. Within the standard four programming languages are defined. Syntax and semantics have been defined, leaving no room for dialects. The languages consist of two textual and two graphical versions:

- Textual,
  Instruction List (IL),
  Structured Text (ST).
- Graphical,
  Ladder Diagram (LD),
  Function Block Diagram (FBD).

The Ladder Diagram has its roots in the USA. It is based on the graphical presentation of Relay Ladder Logic. The Instruction List is its European counterpart. As a textual language, it resembles assembler. The Function Block Diagram is very common to the process industry. It expresses the behavior of functions, function blocks and programs as a set of interconnected graphical blocks, like an electronic circuit diagrams. It looks at a system in terms of the flow of signals between processing elements. The Structured Text is a very powerful language with its roots in Ada, Pascal and "C". It can be used excellently for the definition of complex function blocks, which can be used within any of the other languages. Summarizing, the mayor benefits of IEC 1131 are:

- Standardized languages and standardized programming;
- Increasing software re-usability;
- Ability to use the best language, depending on the application;
- Common platform for understanding control software.

However, the software model and the programming languages described in IEC 1131-3 are mainly applicable on stand alone or loosely coupled PLCs, not on Distributed Control Systems (DCS). For these applications it is necessary to introduce a communication system which is able to handle Distributed Function Blocks. The concept of communication defined in IEC 1131-5 [34] can be used to solve this problem. To exchange data between physically separated devices IEC 1131-5 offers several possibilities, but all of them are quite complicated and without supported synchronization between the function blocks. To enhance the communication facilities in DCSs by using IEC 1131–3 it is necessary to first extend the software model of the standard.

A definition of function blocks suited for DCS application is proposed by the IEC Technical Committee (TC65), Function Blocks. This IEC 1499 draft [35] defines a generic architecture and presents guidelines for the use of Function Blocks in distributed industrial-process measurement and control systems (IPMCSs). The models given in this Standard are intended to be generic; domain independent and extensible to the definition and use of function blocks in other standards.

1995: SP88. The ISA's SP88 committee first began work on the specification in 1989, and soon S88.01 [36] became the fundamental building block of good communication between software vendors providing – and companies looking to implement automated batch-manufacturing solutions.

Within the manufacturing sector, production processes are typically defined as continuous, discrete or batch. Continuous processes are those in which output occurs in a continuous flow. Discrete processes are those in which output results in finite quantities of parts that can be counted. Batch processes are those in which output yields a finite quantity of material. In batch control, the mixing and reacting of materials are crucial to the integrity of production. Products are determined by a recipe that has a name and contains all requisite information for manufacturing the product. This includes ingredients or raw materials needed, the order of process steps through which ingredients must pass, the conditions of each step in the process, and the equipment to be used in the process.

The SP88 committee began with two fundamental work items: models and terminology, which focused on verbal communication standards for batch solutions; and data structures and language guidelines, which addressed the digital communication issues involved. Work on the first item has been completed, and has become an international standard. Efforts on the second item are still ongoing.

SP88 defines three reference models: *process model, physical model and procedural control model.* The mapping of process functionality described in the process model with individual equipment provides the possibility to specify unequivocal elements in the procedural control model. The three models allow a systematic breakdown of a process control system into procedures, operations and phases.

The Process Model defines all processes that lead to the production of finite quantities of material by subjecting quantities of input material to a defined order of processing actions using one or more pieces of equipment.

The Physical Model has seven levels, starting at the top with an enterprise, a site and an area. These three levels are defined by business considerations and are not modeled further in this document. The lower levels of the model refer to specific equipment types and are defined by engineering activities.

A Process Cell is a logical grouping of equipment required for the production of one or more batches. A process cell must include all of the components required to run a specific recipe, since recipes cannot cross process cell boundaries.

A Unit is a collection of associated control modules and/or equipment modules that carry out one or more major processing activities. Units are presumed to operate on only one batch at a time. Units operate relatively independent of one another. Often in process plants there exist similar sets of equipment and it is useful to describe these sets as "classes."

An Equipment Module is an instance of a specific phase. A phase is a general representation of a group of equipment, which performs a sub-function for a unit. An instance of a phase is referred to as an equipment module. Each phase contains attributes common to all equipment modules of that specific phase type. When a phase is added to a unit it becomes an equipment module. An equipment module maps to the equipment phase in the engineered logic via tag addresses. Only one instance of a specific phase may be added to each unit. When an equipment module is required in multiple units, it may be shared between the units.

The Procedural Control Model is defined by means of a hierarchical structure, which can be correlated with the earlier defined process- and physical models. Procedural control directs equipment actions to take place in an ordered sequence. There are four levels to this model:

- The Procedure.
- The Unit Procedure.
- The Operation.
- The Phase.

The utilization of S88.01 modularization brings a degree of efficiency and flexibility that ultimately will reduce the burden on the automation team. Separating the procedural control code from the equipment control code facilitates flexibility. Modularizing the software into reusable phases allows for changes or additions of procedures without ever having to modify and test the existing phase code. A phase is written and validated one time in the life of the equipment. The equipment unit procedure is tested only to ensure that the phases are called in the proper order, greatly simplifying the overall software validation effort.

The related standard SP88.02 (Batch Control Part 2: Data Structures and Language Guidelines), currently in its fourth draft and scheduled for publication next year, describes how data needed for batch control can be stored and exchanged in ways that are independent of the actual hardware or software. It also provides more detail about what a batch recipe actually looks like, an issue that S88.01 does not cover fully. At the moment people tend to use IEC 1131 sequential function charts to describe recipes. IEC 1131 is an excellent standard for programmable controller languages, but was not designed for recipes. S88.02 will attempt to provide a more appropriate method. Plants that are based on batch operations also incorporate some degree of continuous processing, and this too creates problems for S88.01.

4
Outlook

Frost and Sullivan published a new study on the PLC market in January 1995 [37]. Although this study observes the general control and automation market, the relative figures may also be true for the biotechnological industry. According to this study, measured in the number-of-units, the worldwide PLC-market will grow 12.3% from 1993 to the year 2000. The study forecasts that the small PLCs, 30% of the market revenues in 1993, will increase to 35% in 2000, while the revenues for medium and large PLCs will dip from 30% to 26%, over said period. The increased importance of software is also reflected in the figures: software revenues are expected to increase from 9 till 12% over the same period.

Taking into account the rapid developments in hard- and software it is quite difficult to sketch how the next millennium's process control systems will look like. This paper will focus only on two aspects out of a myriad of possible technical topics.

4.1
Integration of Enterprise Resource Planing (ERP) Systems into the Control Level

Integration of enterprise solutions to the control level requires flexibility and openness in tools to solve real world integration problems. The objectives of minimizing special code development making maximum usage of off-the-shelf products and bringing systems rapidly on-line are the goal. One must look beyond the needs of batch execution and select an environment, which allows for quality and maintenance integration.

The integration of ERP into the control systems might include:

- Downloading the production schedule to the control system queues for automatic startup;
- Download specific material reservations;
- Uploading material consumption quantities and lot numbers;
- Uploading semi-finished product quantities and batch numbers;
- Uploading critical product release results that need to follow an order.

In order to further automate the control system integration to handle maintenance and product quality by directly scheduling the control systems, there is a need to:

- Automate maintenance scheduling within ERP for a more refined production schedule;
- Upload up-to-date resource allocation information from the control systems;
- Upload equipment usage from the control systems to proactively schedule maintenance;
- Manage lab samples and map their results to production batches.

Interest in the integration of process control systems with enterprise systems results from the long-term goal to have only one single source of information for entities like bill of materials, inventory, and genealogy. In addition, further advantages would be:

- Minimization of manual data entry errors;
- Minimization of recipe and Bill-Of-Material maintenance;
- Production of complex data summaries not available manually;
- Automated handling of process control system batch scheduling queues;
- Corporate benefits such as accurate planning, reduced inventories, accurate production cost tracking, accurate material tracking, optimal utilization of resources and people, accurate record keeping for quality certification, as well as specification of production targets.

4.2 Ethernet-Based Device-Level Networks

Ethernet TCP/IP based network technology is widely used in office environments to connect everything from mainframe computers to desktop PCs and printers. Although the Ethernet standard used on plant floors is the same as used in office environments, its implementation is considerably different, especially when it comes to redundancy. Several control system manufacturers are using Ethernet to connect controllers and operator interface devices together to form control-centric Ethernet architectures. Pushing nearer to the plant floor, a few manufacturers are forming device-centric Ethernet architectures where I/O devices are directly connected to an Ethernet network or are collected in Ethernet-compatible I/O data concentrators. The use of Ethernet in future projects depends on the application requirements, including how devices receive power and if intrinsic safety is needed. For either of those requirements Ethernet is not well suited. In addition, any factory level control network must be segmented and protected from business traffic to ensure the required reliability and determinism.

4.3 A Trend in the Future?

As pointed out in this review there is a wide gap between industrial requirements and academic research activities. On the academic side, sophisticated self-adapting controllers and complex mechanistic models to observe a process, on the industrial side, simple PID-algorithms, plenty of formalisms and process control based on empirical cultivation techniques. The time should come to merge the two approaches, in order to fulfill all the requirements of future biotechnological processes. The necessary computing performance is now available.

Acknowledgement. The authors would like to thank Chris Beaham for his essential input into this review and his intense editing work.

References

1. GAMP Good Automated Manufacturing Practice Suppliers-Guide Version 3.0, GAMP Forum, March 1998
2. Singh V (1990) In: Omstead DR (ed) Computer Control of Fermentation Processes, p 257
3. Fiechter A, Ingold W, Baerfuss A (1964) Chem Ing Tech 36:1000
4. Proceedings of the 7th IFAC International Conference (1998) Yoshida T, Shioya S (eds)
5. Alford JS (1990) Computer Control of Fermentation Processes, Omstead DR (ed) 221
6. Bader FG, Peterson JA (1990) Computer Control of Fermentation Processes, Omstead DR (ed) 281
7. Reda KD, Omstead RR (1990) Computer Control of Fermentation Processes, Omstead DR (ed) 73
8. Buckland BC (1990) Computer Control of Fermentation Processes, Omstead DR ed.: 237
9. Zabriskie DW, Wareheim DA (1990) Computer Control of Fermentation Processes, Omstead DR (ed) 257
10. Röhr M, Hampel W, Bach HP, Wöhrer W (1978) Rothenburger Symposium: 35
11. Beyeler W, Strub S, Kappel W (1982) 5. Symposium Technische Mikrobiologie, Berlin
12. Gollmer K, Posten C (1995) J Biotechnol 40:99
13. Diehl U, Kappel W, Hass V, Munack A (1990) DECHEMA-Biotechnol. Conf. 4:1003
14. Gregory ME, Keay PJ, Dean P, Bulmer M, Thornbill NF (1993) J Biotechnol 33:223
15. Diaz C, Dieu P, Feuillerat C, Lelong Ph, Salomé M (1995) J Biotechnol 43:21
16. U.S. Department of Health and Human Services, Public Health Services, (1983) "Guide to Inspection of Computerized Systems in Drug Processing", Food and Drug Administration, Washington, DC
17. FDA's Compliance Policy Guide 7132a.07 (1983) "Computer Drug Processing: Input/Output Checking"
18. FDA's Compliance Policy Guide, 7132a.08, Chapter 32a (1983) "Computerized Drug Processing: Identification of 'Persons' on Batch Production and Control Records"
19. FDA's Compliance Policy Guide, 7132a.11, Chapter 32a (1983) "Computerized Drug Processing: CGMP Applicability to Hardware and Software"
20. FDA's Compliance Policy Guide, 7132a.12, Chapter 32a (1985) "Computerized Drug Processing: Vendor Responsibility"
21. FDA's Compliance Policy Guide, 7132a.15, Chapter 32a, (1987) "Computerized Drug Processing: Source Code for Process Control Application Programs"
22. Technical Report Food and Drug Administration, Washington, DC (1987) "Software Development Activities", Reference Materials and Training Aids for Investigators
23. Pharmaceutical Manufacturers Association CSVC (1986), Pharm Technol 10(5):24
24. European Commission (1998), The rules governing medicinal products in the European Union, Volume 4, Good manufacturing practices for medicinal products for human and veterinary use
25. European Commission(1991) Commission Directive 91/356/EEC, The principles and guidelines of good manufacturing practice for medicinal products for human use
26. European Commission(1991) Commission Directive 91/412/EEC, The principles and guidelines of good manufacturing practice for medicinal products for veterinary medicinal products
27. European Commission (1992): The rules governing medicinal products in the European Union, Volume 4. Good manufacturing practices for medicinal products for human and veterinary use, Annex 11: Computerized Systems, Further information: All regulatory documents related to medicinal products in the European Union (Volume 1–Volume 7) can be viewed or downloaded from the "EudraLex homepage" at: http://dg3.eudra.org/ eudralex/index.htm
28. OECD Environment Directorate, Paris (1995) GLP Consensus Document: The Application of the Principles of GLP to Computerized Systems
29. Stokes, Teri: Can You Trust Your Electronic Data? A Tutorial in Computer Validation, DEC White Papers, available at the associated URL address: http://www.digital.com/online/whitepapers/guest05.htm

30. Food and Drug Administration, Washington, D. C (1997).Code of Federal Regulations, Title 21, Food and Drugs, Part 11 (21 CFR Part 11); Electronic Records; Electronic Signatures Further information: The article "Computer Validation: Available Document resources from FDA" provides a list of resources related to computer systems and/or computer validation that are available from the FDA and other US government organizations. The authors have compiled a quite complete list divided into different sections, URL addresses http://www.pharmaportal.com/articles/pt.fda99.cfm
31. International Organization for Standardization (1997) Quality management and quality assurance standards – Part 3: Guidelines for the application of ISO 9001 to the development, supply and maintenance of software, ISO 9000–3: (E)
32. International Organization for Standardization (1994), Quality systems – Model for quality assurance in design, development, production, installation and servicing, ISO 9001: (E)
33. International Electrotechnical Commission (1983), IEC 61131–3: Programmable controllers, Part 3 – Programming languages
34. International Electrotechnical Commission (1998) Draft IEC 61131–5: Programmable controllers, Part 5 – Communications
35. International Electrotechnical Commission (1998), Draft IEC 61299: Function Blocks for industrial-process measurement and control systems
36. Instrument Society of America (1995) Batch Control Part 1: Models and Terminology, ANSI/ISA-S88.01
37. Frost and Sullivan (1995) World programmable logic controller markets

Received September 1999

Author Index Volume 1–70

Author Index Volumes 1–50 see Volume 50

Adam, W., Lazarus, M., Saha-Möller, C. R., Weichhold, O., Hoch, U. Häring, D., Schreier, Ü.: Biotransformations with Peroxidases. Vol. 63, p. 73

Allan, J. V., Roberts, S. M., Williamson, N. M.: Polyamino Acids as Man-Made Catalysts. Vol. 63, p. 125

Al-Rubeai, M.: Apoptosis and Cell Culture Technology. Vol. 59, p. 225

Al-Rubeai, M. see Singh, R. P.: Vol. 62, p. 167

Alsberg, B. K. see Shaw, A. D.: Vol. 66, p. 83

Antranikian, G. see Ladenstein, R.: Vol. 61, p. 37

Antranikian, G. see Müller, R.: Vol. 61, p. 155

Archelas, A. see Orru, R. V. A.: Vol. 63, p. 145

Argyropoulos, D. S.: Lignin. Vol. 57, p. 127

Arnold, F. H., Moore, J. C.: Optimizing Industrial Enzymes by Directed Evolution. Vol. 58, p. 1

Akhtar, M., Blanchette, R. A., Kirk, T. K.: Fungal Delignification and Biochemical Pulping of Wood. Vol. 57, p. 159

Autuori, F., Farrace, M. G., Oliverio, S., Piredda, L., Piacentini, G.: "Tissie" Transglutaminase and Apoptosis. Vol. 62, p. 129

Azerad, R.: Microbial Models for Drug Metabolism. Vol. 63, p. 169

Bajpai, P., Bajpai, P. K.: Realities and Trends in Emzymatic Prebleaching of Kraft Pulp. Vol. 56, p. 1

Bajpai, P. Bajpai, P. K.: Reduction of Organochlorine Compounds in Bleach Plant Effluents. Vol. 57, p. 213

Bajpai, P. K. see Bajpai, P.: Vol. 56, p. 1

Bajpai, P. K. see Bajpai, P.: Vol. 57, p. 213

Bárzana, E.: Gas Phase Biosensors. Vol. 53, p. 1

Bazin, M. J. see Markov, S. A.: Vol. 52, p. 59

Bellgardt, K.-H.: Process Models for Production of β-Lactam Antibiotics. Vol. 60, p. 153

Beppu, T.: Development of Applied Microbiology to Modern Biotechnology in Japan. Vol. 69, p. 41

Berovic, M. see, Mitchell, D. A.: Vol. 68, 61

Beyeler, W., DaPra, E., Schneider, K.: Automation of Industrial Bioprocesses. Vol. 70, p. 139

Beyer, M. see Seidel, G.: Vol. 66, p. 115

Bhatia, P. K., Mukhopadhyay, A.: Protein Glycosylation: Implications for in vivo Functions and Thereapeutic Applications. Vol. 64, p. 155

Bisaria, V. S. see Ghose, T. K.: Vol. 69, p. 87

Blanchette R. A. see Akhtar, M.: Vol. 57, p. 159

de Bont, J.A.M. see van der Werf, M. J.: Vol. 55, p. 147

Bocker, H., Knorre, W. A.: Antibiotica Research in Jena from Penicillin and Nourseothricin to Interferon. Vol. 70, p. 35

Brainard, A. P. see Ho, N. W. Y.: Vol. 65, p. 163

Broadhurst, D. see Shaw, A. D.: Vol. 66, p. 83

Bruckheimer, E. M., Cho, S. H., Sarkiss, M., Herrmann, J., McDonell, T. J.: The Bcl-2 Gene Family and Apoptosis. Vol 62, p. 75
Buchert, J. see Suurnäkki, A.: Vol. 57, p. 261
Bungay, H. R. see Mühlemann, H. M.: Vol. 65, p. 193
Bungay, H.R., Isermann, H.P.: Computer Applications in Bioprocessin. Vol. 70, p. 109

Cao, N. J. see Gong, C. S.: Vol. 65, p. 207
Cao, N. J. see Tsao, G. T.: Vol. 65, p. 243
Carnell, A. J.: Stereoinversions Using Microbial Redox-Reactions. Vol. 63, p. 57
Cen, P., Xia, L.: Production of Cellulase by Solid-State Fermentation. Vol. 65, p. 69
Chang, H. N. see Lee, S. Y.: Vol. 52, p. 27
Cheetham, P. S. J.: Combining the Technical Push and the Business Pull for Natural Flavours.Vol. 55, p. 1
Cho, S. H. see Bruckheimer, E. M.: Vol. 62, p. 75
Chen, Z. see Ho, N. W. Y.: Vol. 65, p. 163 – 192
Christensen, B., Nielsen, J.: Metabolic Network Analysis – A Powerful Tool in Metabolic Engineering. Vol. 66, p. 209
Ciaramella, M. see van der Oost, J.: Vol. 61, p. 87
Contreras, B. see Sablon, E.: Vol. 68, p. 21
Cornet, J.-F., Dussap, C. G., Gros, J.-B.: Kinetics and Energetics of Photosynthetic Micro-Organisms in Photobioreactors. Vol. 59, p. 153
da Costa, M.S., Santos, H., Galinski, E.A.: An Overview of the Role and Diversity of Compatible Solutes in Bacteria and Archaea. Vol. 61, p. 117
Cotter, T. G. see McKenna, S. L.: Vol. 62, p. 1
Croteau, R. see McCaskill, D.: Vol. 55, p. 107

Danielsson, B. see Xie, B.: Vol. 64, p. 1
DaPra, E. see Beyeler, W.: Vol. 70, p. 139
Darzynkiewicz, Z., Traganos, F.: Measurement of Apoptosis. Vol 62, p. 33
Davey, H. M. see Shaw, A. D.: Vol. 66, p. 83
Dean, J. F. D., LaFayette, P. R., Eriksson, K.-E. L., Merkle, S. A.: Forest Tree Biotechnolgy. Vol. 57, p. 1
Demain, A.L., Fang, A.: The natural Functions of Secondary Metabolites. Vol. 69, p. 1
Dochain, D., Perrier, M.: Dynamical Modelling, Analysis, Monitoring and Control Design for Nonlinear Bioprocesses. Vol. 56, p. 147
Du, J. see Gong, C. S: Vol. 65, p. 207 – 241
Du, J. see Tsao, G. T.: Vol. 65, p. 243 – 280
Dueser, M. see Raghavarao, K. S. M. S.: Vol. 68, p. 139
Dussap, C. G. see Cornet J.-F.: Vol. 59, p. 153
Dutta, N. N. see Ghosh, A. C.: Vol. 56, p. 111

Eggeling, L., Sahm, H., de Graaf, A.A.: Quantifying and Directing Metabolite Flux: Application to Amino Acid Overproduction. Vol. 54, p. 1
Ehrlich, H. L. see Rusin, P.: Vol. 52, p. 1
Elias, C. B., Joshi, J. B.: Role of Hydrodynamic Shear on Activity and Structure of Proteins. Vol. 59, p. 47
Elling, L.: Glycobiotechnology: Enzymes for the Synthesis of Nucleotide Sugars. Vol. 58, p. 89
Eriksson, K.-E. L. see Kuhad, R. C.: Vol. 57, p. 45
Eriksson, K.-E. L. see Dean, J. F. D.: Vol. 57, p. 1

Faber, K. see Orru, R. V. A.: Vol. 63, p. 145
Fang, A. see Demain, A. L.: Vol. 69, p. 1
Farrell, R. L., Hata, K., Wall, M. B.: Solving Pitch Problems in Pulp and Paper Processes. Vol. 57, p. 197

Farrace, M. G. see Autuori, F.: Vol. 62, p. 129
Fiechter, A.: Biotechnology in Switzerland and a Glance at Germany. Vol. 69, p. 175
Fiechter, A. see Ochsner, U. A.: Vol. 53, p. 89
Foody, B. see Tolan, J. S.: Vol. 65, p. 41
Freitag, R., Hórvath, C.: Chromatography in the Downstream Processing of Biotechnological Products. Vol. 53, p. 17
Furstoss, R. see Orru, R. V. A.: Vol. 63, p. 145

Galinski, E. A. see da Costa, M. S.: Vol. 61, p. 117
Gatfield, I.L.: Biotechnological Production of Flavour-Active Lactones. Vol. 55, p. 221
Gemeiner, P. see Stefuca, V.: Vol. 64, p. 69
Gerlach, S. R. see Schügerl, K.: Vol. 60, p. 195
Ghose, T.K., Bisaria, V.S.: Development of Biotechnology in India. Vol. 69, p. 71
Ghosh, A. C., Mathur, R. K., Dutta, N. N.: Extraction and Purification of Cephalosporin Antibiotics. Vol. 56, p. 111
Ghosh, P. see Singh, A.: Vol. 51, p. 47
Gilbert, R. J. see Shaw, A. D.: Vol. 66, p. 83
Gomes, J., Menawat, A. S.: Fed-Batch Bioproduction of Spectinomycin. Vol. 59, p. 1
Gong, C. S., Cao, N. J., Du, J., Tsao, G. T.: Ethanol Production from Renewable Resources. Vol. 65, p. 207–241
Gong, C. S. see Tsao, G. T.: Vol. 65, p. 243
Goodacre, R. see Shaw, A. D.: Vol. 66, p. 83
de Graaf, A.A. see Eggeling, L.: Vol. 54, p. 1
de Graaf, A.A. see Weuster-Botz, D.: Vol. 54, p. 75
de Graaf, A.A. see Wiechert, W.: Vol. 54, p. 109
Grabley, S., Thiericke, R.: Bioactive Agents from Natural Sources: Trends in Discovery and Application. Vol. 64, p. 101
Griengl, H. see Johnson, D. V.: Vol. 63, p. 31
Gros, J.-B. see Larroche, C.: Vol. 55, p. 179
Gros, J.-B. see Cornet, J. F.: Vol. 59, p. 153
Guenette M. see Tolan, J. S.: Vol. 57, p. 289
Gutman, A. L., Shapira, M.: Synthetic Applications of Enzymatic Reactions in Organic Solvents. Vol. 52, p. 87

Häring, D. see Adam, E.: Vol. 63, p. 73
Hall, D. O. see Markov, S. A.: Vol. 52, p. 59
Hall, P. see Mosier, N. S.: Vol. 65, p. 23
Harvey, N. L., Kumar, S.: The Role of Caspases in Apoptosis. Vol. 62, p. 107
Hasegawa, S., Shimizu, K.: Noninferior Periodic Operation of Bioreactor Systems. Vol. 51, p. 91
Hata, K. see Farrell, R. L.: Vol. 57, p. 197
Hembach, T. see Ochsner, U. A.: Vol. 53, p. 89
Henzler, H.-J.: Particle Stress in Bioreactor. Vol. 67, p. 35
Herrmann, J. see Bruckheimer, E. M.: Vol. 62, p. 75
Hill, D. C., Wrigley, S. K., Nisbet, L. J.: Novel Screen Methodologies for Identification of New Microbial Metabolites with Pharmacological Activity. Vol. 59, p. 73
Hiroto, M. see Inada, Y.: Vol. 52, p. 129
Ho, N. W. Y., Chen, Z., Brainard, A. P. Sedlak, M.: Successful Design and Development of Genetically Engineering Saccharomyces Yeasts for Effective Cofermentation of Glucose and Xylose from Cellulosic Biomass to Fuel Ethanol. Vol. 65, p. 163
Hoch, U. see Adam, W.: Vol. 63, p. 73
Holló, J, Kralovánsky, U.P.: Biotechnology in Hungary. Vol. 69, p. 151
Hórvath, C. see Freitag, R.: Vol. 53, p. 17
Hummel, W.: New Alcohol Dehydrogenases for the Synthesis of Chiral Compounds. Vol. 58, p.145

Inada, Y., Matsushima, A., Hiroto, M., Nishimura, H., Kodera, Y.: Chemical Modifications of Proteins with Polyethylen Glycols. Vol. 52, p. 129
Iyer, P. see Lee, Y. Y.: Vol. 65, p. 93-115
Irwin, D. C. see Wilson, D. B.: Vol. 65, p. 1
Isermann, H. P. see Bungay, H. R.: Vol. 70, p. 109

Jeffries, T. W., Shi, N.-Q.: Genetic Engineering for Improved Xylose Fementation by Yeasts. Vol. 65, p. 117
Johnson, E. A., Schroeder, W. A.: Microbial Carotenoids. Vol. 53, p. 119
Johnson, D. V., Griengl, H.: Biocatalytic Applications of Hydroxynitrile. Vol. 63, p. 31
Joshi, J. B. see Elias, C. B.: Vol. 59, p. 47
Johnsurd, S. C.: Biotechnolgy for Solving Slime Problems in the Pulp and Paper Industry. Vol. 57, p. 311

Kaderbhai, N. see Shaw, A. D.: Vol. 66, p. 83
Kataoka, M. see Shimizu, S.: Vol. 58, p. 45
Kataoka, M. see Shimizu, S.: Vol. 63, p. 109
Katzen, R., Tsao, G. T.: A View of the History of Biochemical Engineering. Vol. 70, p. 77
Kawai, F.: Breakdown of Plastics and Polymers by Microorganisms. Vol. 52, p. 151
Kell, D. B. see Shaw, A. D.: Vol. 66, p. 83
Kieran, P. M., Malone, D. M., MacLoughlin, P. F.: Effects of Hydrodynamic and Interfacial Forces on Plant Cell Suspension Systems. Vol. 67, p. 139
King, R.: Mathematical Modelling of the Morphology of Streptomyces Species. Vol. 60, p. 95
Kirk, T. K. see Akhtar, M.: Vol. 57, p. 159
Knorre, W. A. see Bocker, H.: Vol. 70, p. 35
Kobayashi, M. see Shimizu, S.: Vol. 58, p. 45
Kodera, F. see Inada, Y.: Vol. 52, p. 129
Kossen, N. W. F.: The Morphology of Filamentous Fungi. Vol. 70, p. 1
Krabben, P. Nielsen, J.: Modeling the Mycelium Morphology of Penicilium Species in Submerged Cultures. Vol. 60, p. 125
Kralovánszky, U. P. see Holló, J.: Vol. 69, p. 151
Krämer, R.: Analysis and Modeling of Substrate Uptake and Product Release by Procaryotic and Eucaryotik Cells. Vol. 54, p. 31
Kretzmer G.: Influence of Stress on Adherent Cells. Vol. 67, p. 123
Krieger, N. see Mitchell, D. A.: Vol. 68, p. 61
Kuhad, R. C., Singh, A., Eriksson, K.-E. L.: Microorganisms and Enzymes Involved in the Degradation of Plant Cell Walls. Vol. 57, p. 45
Kuhad, R. Ch. see Singh, A.: Vol. 51, p. 47
Kumagai, H.: Microbial Production of Amino Acids in Japan. Vol. 69, p. 71
Kumar, S. see Harvey, N. L.: Vol. 62, p. 107

Ladenstein, R., Antranikian, G.: Proteins from Hyperthermophiles: Stability and Enzamatic Catalysis Close to the Boiling Point of Water. Vol. 61, p. 37
Ladisch, C. M. see Mosier, N. S.: Vol. 65, p. 23
Ladisch, R. M. see Mosier, N. S.: Vol. 65, p. 23
Lammers, F., Scheper, T.: Thermal Biosensors in Biotechnology. Vol. 64, p. 35
Larroche, C., Gros, J.-B.: Special Transformation Processes Using Fungal Spares and Immobilized Cells. Vol. 55, p. 179
LaFayette, P. R. see Dean, J. F. D.: Vol. 57, p. 1
Lazarus, M. see Adam, W.: Vol. 63, p. 73
Leak, D. J. see van der Werf, M. J.: Vol. 55, p. 147
Lee, S. Y., Chang, H. N.: Production of Poly(hydroxyalkanoic Acid). Vol. 52, p. 27
Lee, Y. Y., Iyer, P., Torget, R. W.: Dilute-Acid Hydrolysis of Lignocellulosic Biomass. Vol. 65, p. 93
Lievense, L. C., van't Riet, K.: Convective Drying of Bacteria II. Factors Influencing Survival. Vol. 51, p. 71

Mac Loughlin, P.F. see Kieran, P. M.: Vol. 67, p. 139
Malone, D. M. see Kieran, P. M.: Vol. 67, p. 139
Maloney, S. see Müller, R.: Vol. 61, p. 155
Mandenius, C.-F.: Electronic Noses for Bioreactor Monitoring. Vol. 66, p. 65
Markov, S. A., Bazin, M. J., Hall, D. O.: The Potential of Using Cyanobacteria in Photobioreactors for Hydrogen Production. Vol. 52, p. 59
Marteinsson, V. T. see Prieur, D.: Vol. 61, p. 23
Mathur, R. K. see Ghosh, A. C.: Vol. 56, p. 111
Matsushima, A. see Inada, Y.: Vol. 52, p. 129
McCaskill, D., Croteau, R.: Prospects for the Bioengineering of Isoprenoid Biosynthesis. Vol. 55, p. 107
McDonell, T. J. see Bruckheimer, E. M.: Vol. 62, p. 75
McGovern, A. see Shaw, A. D.: Vol. 66, p. 83
McGowan, A. J. see McKenna, S. L.: Vol. 62, p. 1
McKenna, S. L.: McGowan, A. J., Cotter, T. G.: Molecular Mechanisms of Programmed Cell Death. Vol. 62, p. 1
McLoughlin, A. J.: Controlled Release of Immobilized Cells as a Strategy to Regulate Ecological Competence of Inocula. Vol. 51, p. 1
Menachem, S. B. see Argyropoulos, D. S. : Vol. 57, p. 127
Menawat, A. S. see Gomes J.: Vol. 59, p. 1
Menge, M. see Mukerjee, J.: Vol. 68, p. 1
Merkle, S. A. see Dean, J. F. D.: Vol. 57, p. 1
Mitchell, D.A., Berovic, M., Krieger, N.: Biochemical Engineering Aspects of Solid State Bioprocessing. Vol. 68, p. 61
Moore, J. C. see Arnold, F. H.: Vol. 58, p. 1
Mosier, N.S., Hall, P., Ladisch, C.M., Ladisch, M.R.: Reaction Kinetics, Molecular Action, and Mechanisms of Cellulolytic Proteins. Vol. 65, p. 23
Moracci, M. see van der Oost, J.: Vol. 61, p. 87
Mühlemann, H.M. Bungay, H.R.: Research Perspectives for Bioconversion of Scrap Paper. Vol. 65, p. 193
Müller, R., Antranikian, G., Maloney, S., Sharp, R.: Thermophilic Degradation of Environmental Pollutants. Vol. 61, p. 155
Mukherjee, J., Menge, M.: Progress and Prospects of Ergot Alkaloid Research. Vol. 68, p. 1
Mukhopadhyay, A.: Inclusion Bodies and Purification of Proteins in Biologically Active Forms. Vol. 56, p. 61
Mukhopadhyay, A. see Bhatia, P.K.: Vol. 64, p. 155

Nielsen, J. see Christensen, B.: Vol. 66, p. 209
Nielsen, J. see Krabben, P.: Vol. 60, p. 125
Nisbet, L.J. see Hill, D.C.: Vol. 59, p. 73
Nishimura, H. see Inada, Y.: Vol. 52, p. 123

Ochsner, U.A., Hembach, T., Fiechter, A.: Produktion of Rhamnolipid Biosurfactants. Vol. 53, p. 89
O'Connor, R.: Survival Factors and Apoptosis: Vol. 62, p. 137
Ogawa, J. see Shimizu, S.: Vol. 58, p. 45
Ohta, H.: Biocatalytic Asymmetric Decarboxylation. Vol. 63, p. 1
van der Oost, J., Ciaramella, M., Moracci, M., Pisani, F.M., Rossi, M., de Vos, W.M.: Molecular Biology of Hyperthermophilic Archaea. Vol. 61, p. 87
Oliverio, S. see Autuori, F.: Vol. 62, p. 129
Orru, R. V.A., Archelas, A., Furstoss, R., Faber, K.: Epoxide Hydrolases and Their Synthetic Applications. Vol. 63, p. 145

Paul, G.C., Thomas, C.R.: Characterisation of Mycelial Morphology Using Image Analysis. Vol. 60, p. 1
Perrier, M. see Dochain, D.: Vol. 56, p. 147

Piacentini, G. see Autuori, F.: Vol. 62, p. 129
Piredda, L. see Autuori, F.: Vol. 62, p. 129
Pisani, F. M. see van der Oost, J.: Vol. 61, p. 87
Pohl, M.: Protein Design on Pyruvate Decarboxylase (PDC) by Site-Directed Mutagenesis. Vol. 58, p. 15
Pons, M.-N., Vivier, H.: Beyond Filamentous Species. Vol. 60, p. 61
Pons, M.-N., Vivier, H.: Biomass Quantification by Image Analysis. Vol. 66, p. 133
Prieur, D., Marteinsson, V. T.: Prokaryotes Living Under Elevated Hydrostatic Pressure. Vol. 61, p. 23
Pulz, O., Scheibenbogen, K.: Photobioreactors: Design and Performance with Respect to Light Energy Input. Vol. 59, p. 123

Raghavarao, K. S. M. S., Dueser, M., Todd, P.: Multistage Magnetic and Electrophoretic Extraction of Cells, Particles and Macromolecules. Vol. 68, p. 139
Ramanathan, K. see Xie, B.: Vol. 64, p. 1
van't Riet, K. see Lievense, L. C.: Vol. 51, p. 71
Roberts, S. M. see Allan, J. V.: Vol. 63, p. 125
Roehr, M.: History of Biotechnology in Austria. Vol. 69, p. 125
Rogers, P. L., Shin, H. S., Wang, B.: Biotransformation for L-Ephedrine Production. Vol. 56, p. 33
Rossi, M. see van der Oost, J.: Vol. 61, p. 87
Roychoudhury, P. K., Srivastava, A., Sahai, V.: Extractive Bioconversion of Lactic Acid. Vol. 53, p. 61
Rowland, J. J. see Shaw, A. D.: Vol. 66, p. 83
Rusin, P., Ehrlich, H. L.: Developments in Microbial Leaching – Mechanisms of Manganese Solubilization. Vol. 52, p. 1
Russell, N. J.: Molecular Adaptations in Psychrophilic Bacteria: Potential for Biotechnological Applications. Vol. 61, p. 1

Sablon, E., Contreras, B., Vandamme, E.: Antimicrobial Peptides of Lactic Acid Bacteria: Mode of Action, Genetics and Biosynthesis. Vol. 68, p. 21
Sahai, V. see Singh, A.: Vol. 51, p. 47
Sahai, V. see Roychoudhury, P. K.: Vol. 53, p. 61
Saha-Möller, C. R. see Adam, W.: Vol. 63, p. 73
Sahm, H. see Eggeling, L.: Vol. 54, p. 1
Saleemuddin, M.: Bioaffinity Based Immobilization of Enzymes. Vol. 64, p. 203
Santos, H. see da Costa, M. S.: Vol. 61, p. 117
Sarkiss, M. see Bruckheimer, E. M.: Vol. 62, p. 75
Scheibenbogen, K. see Pulz, O.: Vol. 59, p. 123
Scheper, T. see Lammers, F.: Vol. 64, p. 35
Schneider, K. see Beyeler, W.: Vol. 70, p. 139
Schreier, P.: Enzymes and Flavour Biotechnology. Vol. 55, p. 51
Schreier, P. see Adam, W.: Vol. 63, p. 73
Schroeder, W. A. see Johnson, E. A.: Vol. 53, p. 119
Schügerl, K., Gerlach, S. R., Siedenberg, D.: Influence of the Process Parameters on the Morphology and Enzyme Production of Aspergilli. Vol. 60, p. 195
Schügerl, K. see Seidel, G.: Vol. 66, p. 115
Schügerl, K.: Recovery of Proteins and Microorganisms from Cultivation Media by Foam Flotation. Vol. 68, p. 191
Schügerl, K.: Development of Bioreaction Engineering. Vol. 70, p. 41
Schumann, W.: Function and Regulation of Temperature-Inducible Bacterial Proteins on the Cellular Metabolism. Vol. 67. p. 1
Schuster, K. C.: Monitoring the Physiological Status in Bioprocesses on the Cellular Level. Vol. 66, p. 185
Scouroumounis, G. K. see Winterhalter, P.: Vol. 55, p. 73
Scragg, A.H.: The Production of Aromas by Plant Cell Cultures. Vol. 55, p. 239

Sedlak, M. see Ho, N. W. Y.: Vol. 65, p. 163
Seidel, G., Tollnick, C., Beyer, M., Schügerl, K.: On-line and Off-line Monitoring of the Production of Cephalosporin C by Acremonium Chrysogenum. Vol. 66, p. 115
Shamlou, P. A. see Yim, S. S.: Vol. 67. p. 83
Shapira, M. see Gutman, A. L.: Vol 52, p. 87
Sharp, R. see Müller, R.: Vol. 61, p. 155
Shaw, A. D., Winson, M. K., Woodward, A. M., McGovern, A., Davey, H. M., Kaderbhai, N., Broadhurst, D., Gilbert, R. J., Taylor, J., Timmins, E. M., Alsberg, B. K., Rowland, J. J., Goodacre, R., Kell, D. B.: Rapid Analysis of High-Dimensional Bioprocesses Using Multivariate Spectroscopies and Advanced Chemometrics. Vol. 66, p. 83
Shi, N.-Q. see Jeffries, T. W.: Vol. 65, p. 117
Shimizu, S., Ogawa, J., Kataoka, M., Kobayashi, M.: Screening of Novel Microbial for the Enzymes Production of Biologically and Chemically Useful Compounds. Vol. 58, p. 45
Shimizu, K. see Hasegawa, S.: Vol. 51, p. 91
Shimizu, S., Kataoka, M.: Production of Chiral C3- and C4-Units by Microbial Enzymes. Vol. 63, p. 109
Shin, H. S. see Rogers, P. L.: Vol. 56, p. 33
Siedenberg, D. see Schügerl, K.: Vol. 60, p. 195
Singh, R. P., Al-Rubeai, M.: Apoptosis and Bioprocess Technology. Vol 62, p. 167
Singh, A., Kuhad, R. Ch., Sahai, V., Ghosh, P.: Evaluation of Biomass. Vol. 51, p. 47
Singh, A. see Kuhad, R. C.: Vol. 57, p. 45
Sonnleitner, B.: New Concepts for Quantitative Bioprocess Research and Development. Vol. 54, p. 155
Sonnleitner, B.: Instrumentation of Biotechnological Processes. Vol. 66, p. 1
Stefuca, V., Gemeiner, P.: Investigation of Catalytic Properties of Immobilized Enzymes and Cells by Flow Microcalorimetry. Vol. 64, p. 69
Srivastava, A. see Roychoudhury, P. K.: Vol. 53, p. 61
Suurnäkki, A., Tenkanen, M., Buchert, J., Viikari, L.: Hemicellulases in the Bleaching of Chemical Pulp. Vol. 57, p. 261

Taylor, J. see Shaw, A. D.: Vol. 66, p. 83
Tenkanen, M. see Suurnäkki, A.: Vol. 57, p. 261
Thiericke, R. see Grabely, S.: Vol. 64, p. 101
Thömmes, J.: Fluidized Bed Adsorption as a Primary Recovery Step in Protein Purification. Vol. 58, p. 185
Thomas, C. R. see Paul, G. C.: Vol. 60, p. 1
Timmens, E. M. see Shaw, A. D.: Vol. 66, p. 83
Todd, P. see Raghavarao, K. S. M. S.: Vol. 68, p. 139
Tolan, J. S., Guenette, M.: Using Enzymes in Pulp Bleaching: Mill Applications.Vol. 57, p. 289
Tolan, J. S., Foody, B.: Cellulase from Submerged Fermentation. Vol. 65, p. 41
Tollnick, C. see Seidel, G.: Vol. 66, p. 115
Traganos, F. see Darzynkiewicz, Z.: Vol. 62, p. 33
Torget, R. W. see Lee, Y. Y.: Vol. 65, p. 93-115
Tsao, G. T., Cao, N. J. Du, J., Gong, C. S.: Production of Multifunctional Organic Acids from Renewable Resources. Vol. 65, p. 243
Tsao, G. T. see Gong, C. S.: Vol. 65, p. 207
Tsao, G. T. see Katzen, R.: Vol. 70, p. 77

Vandamme, E. see Sablon, E.: Vol. 68, p. 21
Viikari, L. see Suurnäkki, A.: Vol. 57, p. 261
Vivier, H. see Pons, M.-N.: Vol. 60, p. 61
Vivier, H. see Pons, M.-N.: Vol. 66, p. 133
de Vos, W. M. see van der Oost, J.: Vol. 61, p. 87

Wang, B. see Rogers, P. L.: Vol. 56, p. 33
Wall, M. B. see Farrell, R. L.: Vol. 57, p. 197
Weichold, O. see Adam, W.: Vol. 63, p. 73
van der Werf, M. J., de Bont, J. A. M. Leak, D. J.: Opportunities in Microbial Biotransformation of Monoterpenes. Vol. 55, p. 147
Weuster-Botz, D., de Graaf, A.A.: Reaction Engineering Methods to Study Intracellular Metabolite Concentrations. Vol. 54, p. 75
Wiechert, W., de Graaf, A.A.: In Vivo Stationary Flux Analysis by ^{13}C-Labeling Experiments. Vol. 54, p. 109
Wiesmann, U.: Biological Nitrogen Removal from Wastewater. Vol. 51, p. 113
Williamson, N. M. see Allan, J. V.: Vol. 63, p. 125
Wilson, D. B., Irwin, D. C.: Genetics and Properties of Cellulases. Vol. 65, p. 1
Winson, M. K. see Shaw, A. D.: Vol. 66, p. 83
Winterhalter, P., Skouroumounis, G. K.: Glycoconjugated Aroma Compounds: Occurence, Role and Biotechnological Transformation. Vol. 55, p. 73
Woodley, J.M.: Advances in Enzyme Technology – UK Contributions. Vol. 70, p. 93
Woodward, A. M. see Shaw, A. D.: Vol. 66, p. 83
Wrigley, S. K. see Hill, D. C.: Vol. 59, p. 73

Xia, L. see Cen, P.: Vol. 65, p. 69
Xie, B., Ramanathan, K., Danielsson, B.: Principles of Enzyme Thermistor Systems: Applications to Biomedical and Other Measurements. Vol. 64, p. 1

Yim, S. S., Shamlou, P. A.: The Engineering Effects of Fluids Flow and Freely Suspended Biological Macro-Materials and Macromolecules. Vol. 67. p. 83

Subject Index

Academy of Sciences (GDR) 36, 40
Acid hydrolysis 80, 81
Actinomycin 39
Actuators 144, 145, 150
Airlift-tower-loop reactors 47, 53
All-glass bacterial filters 35
6-Aminopenicillanic acid 99
Analog computers 111
Analysis 50
Animal cell cultures 62
Animal cells, anchorage-dependent 63
Apple computer 115
Applet 132
Arsenic 36
Artificial intelligence 59, 111, 126, 127, 143, 147
Artificial neural networks (ANN) 59
Aspen 120, 125
Aspergillus awamori 11, 20
Aspergillus niger 28
Autographa californica 44
Automation 119, 122
Autoselect System 38
Auxostat 124

Balances, macroscopic 48
BASIC 110, 114–117, 134
Batch 45
BCG-vaccination 36
Beutenberg Campus Jena 36, 40
BHK cells 63
Biomass conversion 77, 80
Bioprocess technology 99
Biosensors 49
Bistability 38
Borosilicatglass 36
Branches 5
Branching15
Bubble-column reactors 47, 53
Bubbles, coalescence 27

Carl Zeiss Jena 35, 36, 38
Catharanthus roseus 44
Cell fermentation 96
Cell models, structured 51
Cell production 95
Cell recycle reactor 62
Cell wall synthesis 12
Cellulase 82, 84, 88
Central Institute of Microbiology and Experimental Therapy 37, 40
Chitin 11, 12
Commodore Amiga 116, 117
Compact stirred-loop reactors 54
Compartment models 52
Computational fluid dynamics (CFD) 48
Computer interfacing 111, 114, 129
Continuous reactors 45
Control 45, 119, 122
Cross-flow-membrane module 62
Cube root law 16
Customer Service 131
Cybernetic modeling 51

Deep-stream fermentors 39
Degree of reduction 48
Demixing 27
DHCD 102
Diffusion, intraparticle 60
Digitalis lanata 44
Disk drive 116
Distributed Control Systems (DCS) 158
Downstream processing 135

Energy balances 45
Enterprise resource planing 161
Environmental conditions 6
Enzyme immobilisation 93, 98
Enzyme isolation 93, 97
Enzyme production 95
Enzyme technology 93
Equipment module 160
Erythromycin 39
Escherichia coli 54, 96
Ethanol production 85, 87
Ethanol recovery 87

Ethernet 162
Expert systems 41, 127
Explosion pretreatment 88
Extended Kalman filter 59

Fat interesterification 101
Fed-batch 45
Fermentation, solid phase 88
–, submerged 78
Fieldbus 143, 150
Flocculation, dynamic 27
Flock 5
Flow injection analysis 51
Fluid dynamics 41, 45, 46
Flux balancing 58
Food and Drug Administration (FDA) 155
Fortran 110, 113
Fourier transform, near-infrared 51
Fragmentation 15
Fuzzy logic 127, 128
Fuzzy reasoning techniques 59

Gamp-forum 143, 156
Genetic Engineering Pilot Plant 40
Genetics 6, 25
GenSym 126
German Academy of Sciences 36
Glucan 11, 13
Good Manufacturing Practice (GMP) 155
Griseufulvin 39
Growth 15

Hans-Knöll-Institute for Natural Products Research 40
High cell density bioprocess 39
High density cultures 62
HKI 40
Hollow fiber reactors 62
HPLC 51
Human-Machine-Interface 142, 148, 149
Hybrid computing 112
Hybrid model 121
Hybridoma cells 44
Hyphae 5
Hyphal element 5

Image analysis 8
Information management 130
Institute of Experimental Microbiology Jena 36
Institute of Molecular Biology Jena 40
Institute of Molecular Biotechnology Jena 40
Institute of Microbiology and Experimental Therapy 36
Institute of Virology Jena 40
Instrumentation 119
Intelligent systems 126
Interferon alpha 1/2 39
Interferon gamma 39
ISO 153, 156

Java 110, 132
Jenaer Glaswerk Schott & Gen. 35, 36
Jenapharm 36

Kinetic equations 45
Kolmogorov 17
Kolmogorov-length-scale 63

Lactobacilli 62
Ladder diagramm 158
LaTeX 115, 116
Life-cycle 155, 157
Lignocellulosics 80
Limiting nutrient 118
Line printer 113, 117
L-Lysin 39

Macintosh 116
Macro-mixer 56
Mammalian cells 63
Mass and energy balances 41
Mass transport/transfer 6, 27
Master/slave 134
Mathematical models 41
Meatbolic pathways 64
Medium choice 101
Metabolic control analysis (MCA) 57
Metabolic engineering 41, 57, 64, 84, 130
Metabolic flux 57
– – analysis 57, 64
Metabolic pathways 49
Michaelis-Menten kinetics 48
Micro-electrodes 10
Micro-tubules 12
Microcapsule bioreactors 63
Microcarriers 62, 63
Microorganisms, immobilized 60
Minicomputer 110, 113, 114, 116
Mixing processes 46
Models 4, 10, 21, 120
–, extrapolation 22
–, physiology based 18
–, robustness 23
–, unstructured 121
Monoclonal antibodies 44
Monod equation 49, 118, 121, 135
Morphology 5, 6

–, implementation in industry 28
–, modelling 14
–, overcoming problems 29
–, production 20
Multiphase reactors 46

Navier-Stokes Differential-Equation System 48
Neural networks 128
Neuraminic acid aldolase 97
Neurospora crassa 25
NMR spectroscopy 57, 58
– –, in-vivo 64
Nourseothricin 39
Nystation 39

Observer 123
On-line sampling 50
Organic acids 88
Oxygen analyzer 86
Oxygen transfer 78, 86
Oxytetracyclin 39

Parameter estimation/optimization 23
Parameter-adaptive techniques 59
Paromomycin 39
Pellet 5
Penicillin 35, 36, 38
Penicillin acylase 99
Penicillium chrysogenum 11, 16–18, 21
Penicillium notatum 35
Personal Computer (PC) 149
pH measurement 49
Phosphat regulating effect 39
Physical model 159
Physiology 11
Physiology based models 18
PID-operation 142, 146, 160
Plant cell cultures 62
Plant cells 63
Plasmid host system 44
Plasmid stability 61
Power consumption 46
Preconditioning 50
Process cell 159
Process control 22
Process design 103, 120
Process identification 57
Process integration 103
Process model 159
Process monitoring 41, 45
Process optimization 22
Programmable logic controller (PLC) 147, 148, 158
Protein engineering 93, 104
Proteolytic enzyme complex 39
Pseudomonas putida 103
Pumps 133
Punch card 113

Reactor choice 100
Reactors 46, 47, 53, 54, 60
Respiratory quotient 56
Responses 129
Resting cells 95
Rhizopus nigricans 94
Rotating-membrane filter 62

Saccharification/fermentation 86
Saccharomyces cerevisiae 25, 54
Sample conditioning 50
Satisfaction of curiosity 22
Scale-up 22, 23
Scaling 111
Schott-Zeiss-Institute 36
SCP 43, 44
Sensors 123, 144, 145, 150
Shear forces 63
Shells 127, 129
Simulation 117, 124
Single cell protein (SCP) 43
Single hyphal elements 9
Solid particles, interactions 27
Soviet Army administration 36
SP88 159
Specific power input (P/V) 53
Spectrum (2n+1) 38
Staining 9
Standard operating procedure 142
Staphylokinase 39
Steady state 121, 129, 136
Stepping motors 133
Steroid conversions 60, 100
Stirred-tank reactors 46, 53
Streptikinase 39
Streptomyces 18, 20
Streptomyces niveus 53
Streptomycin 36, 38, 39
Structured models 121
Sucrose inversion 60

Teaching 132
Teletype 115
Transketolase 96
Transport mechanisms 11
Transport processes 6, 46, 53
Trichosporon cutaneum 54, 56
Tuberculosis 35, 36
Turbidostat 124
Turbulent forces 45

Turbulent jet-flow 63
Turimycin 39

Unit 160
US-Army administration 36

Vaccine production 64
Vaccines 62
Validation 141-144, 153, 155, 157
VEB Arzneimittelwerk Dresden 40
VEB Carl Zeiss Jena 38
VEB Gärungschemie Dessau 39
VEB Jenapharm Jena 36, 38
Vesicles 11
Viscosity 7, 26
- measurements 27

Whole broth properties 26

Xanthomonas campestris 54, 56
Xylose conversion 84

Zanamavir 97
Zentralinstitut für Mikrobiologie und experimentelle Therapie Jena 40
ZIMET 37, 40

Printing: Saladruck, Berlin
Binding: Lüderitz & Bauer, Berlin